CELL MEMBRANES AND
GLYCOPROTEIN SYNTHESIS

CELL MEMBRANES AND GLYCOPROTEIN SYNTHESIS

PROCEEDINGS OF
A ROYAL SOCIETY DISCUSSION MEETING
HELD ON 27 AND 28 MAY 1982

ORGANIZED AND EDITED BY
D. H. NORTHCOTE, F.R.S.

INCLUDING A BRIEF REPORT OF THE
CIBA FOUNDATION MEETING HELD ON 26 MAY 1982

LONDON
THE ROYAL SOCIETY
1983

Printed in Great Britain for the Royal Society
at the
University Press, Cambridge

ISBN 0 85403 200 2

First published in *Philosophical Transactions of the Royal Society of London*,
series B, volume 300 (no. 1099), pages 107–235 (1982).

Published by the Royal Society
6 Carlton House Terrace, London SW1Y 5AG

PREFACE

The papers presented at this Discussion Meeting were concerned with the biosynthesis and function of the oligosaccharide prosthetic groups of glycoproteins found at the membranes of eukaryotic cells. In particular most of the papers dealt with the groups of oligosaccharides that consist in part of a branched chain of mannosyl residues attached through N,N'-diacetyl chitobiose by an N-glycosidic bond to the asparagine residue of the peptide chains.

These oligosaccharide chains in their final form, on different glycoproteins, have diverse structures, although they are formed from a common precursor that is assembled on dolichyl phosphate and consists of $Glc_3Man_9GlcNAc_2$. This branched oligosaccharide is transferred to an asparagine residue of the polypeptide and is subsequently processed by the removal of glucose and some mannose units and extended by further transglycosylase reactions. The enzymic steps involved in the assembly of the oligosaccharide, its transfer to protein and its processing all take place at definite sites within the membrane system of the cell and are control points for glycoprotein synthesis, passage in the membrane system and secretion from the cell.

Detailed knowledge of these synthetic systems has been considerably extended over the last few years. The increasing awareness of the importance of the glycoproteins in many biological processes, including the surface recognition sites of cells, has made the papers collected in this volume a valuable summary of the present knowledge of the subject. The research described was carried out in some of the major laboratories in Europe and the U.S.A. and is still being actively pursued so that in most cases the papers indicate in a very clear way how the subject will be extended in the immediate future.

October 1982

D. H. NORTHCOTE

CONTENTS

[Three plates]

Phil. Trans. R. Soc. Lond. B **300**, 109–115 (1982)
Printed in Great Britain

Report of the Ciba Foundation discussion meeting on protein and membrane interaction held at the Ciba Foundation on 26 May 1982

By W. T. Dixon

*Department of Biochemistry, University of Cambridge, Tennis Court Road,
Cambridge CB2 1QW, U.K.*

The Deputy Director of the Ciba Foundation, Dr Ruth Porter, in her opening remarks, described the main function of the Ciba Foundation as one of encouraging multidisciplinary international cooperation in scientific and medical research. The purpose of this informal half-day discussion meeting was firstly to provide an atmosphere in which participants in the Royal Society Discussion Meeting to be held on the following two days could meet one another, and secondly to hear three lectures loosely grouped around the theme of 'Protein and membrane interaction'.

Professor D. H. Northcote, the organizer of the Royal Society meeting, after welcoming the participants, introduced Dr Gordon Koch of the M.R.C. Molecular Biology Unit, Cambridge. Dr Koch addressed the problem of 'Associations between extracellular glycoproteins and microfilaments in the pseudopodial reactions of motile cells'.

Various morphological specializations of the cell surface are known including pseudopodia, microvilli and pits. Associated with all of these specializations is a local differentiation of the cell surface so that particular surface components become highly enriched in these areas. These local differentiations are related to the function carried out by the morphological specializations, which, in pseudopodial reactions, include locomotion, phagocytosis and cell–cell interactions. It is therefore important to elucidate the basic mechanism that generates the morphological specialization, bearing in mind that the plasma membrane is an intrinsically fluid structure.

In pseudopodia there was a variety of circumstantial evidence to suggest that interactions between the surface and the intracellular microfilaments operate during the function of these organelles.

Koch and co-workers became interested in this problem from the discovery of a cultured cell line, P815, in which the cells possessed a large number of long thin pseudopodia called filopodia. A great advantage of using these cells was that the filopodia could be easily sheared away from the cell body with little or no damage. This allowed the isolation of relatively pure preparations of intact filopodia, which could be subjected to biochemical analysis. In earlier work (Koch & Smith 1978), they had shown that certain surface histocompatibility antigens could form a stable association with the microfilaments from such filopodial preparations. However, they were unable to demonstrate any detectable biological function of such an association.

From a P815 cell suspension culture a variant was isolated that underwent cell spreading when transferred to a substratum. This spreading phenomenon could be readily assayed and served as a laboratory analogue of the pseudopodial reactions. It occurred in two stages whereby the cell would initially produce numerous filopodial extensions, followed by the flow of

[1]

membraneous lamellae between them to adopt the fully spread morphology. A crucial stage of this spreading reaction was the formation of contact between the filopodia and substratum, which occurred via a layer of serum proteins adsorbed to the inert substratum. In fibroblasts this spreading factor is known to be fibronectin. However, depletion of fibronectin from serum did not reduce the capacity of serum to support the spreading activity of the P815 cells, indicating that fibronectin was not the major spreading factor in calf serum.

By using either whole cells or isolated filopodia as immunogens, antibodies were developed to isolate the spreading factor from the growth medium, on the rationale that this factor must have the capacity to bind to the filopodia. Such antisera recognize two major Concanavalin A-binding glycoproteins from calf serum, with molecular masses of 85 and 160 kDa. To determine whether these were responsible for cell spreading, Koch and co-workers adopted two approaches. Using anti-filopodia antibodies, they either specifically depleted the growth medium of the glycoproteins or alternatively overlayered the serum coated substratum with the antibody. In both cases the spreading reaction was inhibited. When they attempted to isolate and purify the individual serum factors, named GP85 and GP160, they were only able to purify the former to homogeneity.

To investigate whether GP85 had any stable interaction with the filopodia, two-dimensional glycoprotein 'fingerprint' gels of whole cells, isolated membranes and whole filopodia were compared. They found that the GP85 was quantitatively enriched in the filopodial preparation. Indeed, when anti-GP85 was made immunofluorescent and applied to P815 cells, preferential staining in 'plaque' regions of intercalation and hence local enrichment of filopodia was found.

Koch and co-workers then tested by the 'myosin-affinity technique' whether the GP85 formed a similar association with microfilaments as they had found earlier with the H-2 antigens. Pure preparations of filopodia were treated with detergent to dissolve membranes and combined with synthetic filaments of rabbit muscle myosin to selectively extract actin, the main component of microfilaments. The remainder fraction and the spun-out microfilament fraction were analysed for the retention of the GP85 and GP160. The microfilament fraction was found to be highly enriched with both of the serum glycoproteins.

This suggested a direct physical connection between the two entities, but the problem still remained of identifying the GP85 receptor on the filopodia. A clue to this problem had arisen when cell lines similar to P815 were infected with RNA tumour viruses. These shed large quantities of virus particles rather than filopodia but interestingly were also found to contain high concentrations of GP85. The receptor might therefore be a viral surface component, and subsequent electron-microscope examination of whole cells revealed single dense RNA tumour virus particles at the tips of filopodia. This surface component might be involved in the attachment of the serum spreading factors to the surface of the filopodia and it was probably the envelope glycoprotein of the virus. To test this, a panel of well characterized polyclonal and monoclonal antibodies to the major GP70 viral glycoprotein were used. After surface labelling P815 cells the anti-GP70 antibodies immunoprecipitated a single component with a molecular mass of 85 kDa. In control experiments on cells grown in the absence of calf serum the immunoprecipitable glycoprotein was absent. This suggested that the antiGP70 was pulling down the GP85 serum protein and two-dimensional gels of the 85 kDa glycoprotein supported this identification.

In summary, Koch and co-workers have shown not only an association between one of the major serum spreading factors and the filopodia but also that these serum factors are pre-

[2]

ferentially retained by the microfilaments. In addition, evidence was presented for an association between a viral GP70 antigen and the filopodia, during spreading reactions and that this may be the receptor for the GP85 serum factor.

The second speaker, Dr Graham Warren of E.M.B.L. in Heidelberg, Germany, presented the work that he and his colleagues have performed on the 'Structural and functional dissection of the Golgi complex'. They have used an enveloped-animal virus, Semliki Forest virus, as a model system for plasma membrane proteins that are assembled in the endoplasmic reticulum and pass through the Golgi apparatus on their way to the cell surface.

The virus consists of a central nucleocapsid composed of a single RNA molecule complexed with about 240 capsid proteins and surrounded by a host-derived membrane bilayer. There are about 240 spike proteins in the membrane, each composed of three component glycoprotein molecules, E_1, E_2 and E_3. The first two of these are membrane-spanning proteins, and the third is peripheral. Both E_2 and E_3 are synthesized as a precursor in the endoplasmic reticulum and are later cleaved. These glycoproteins carry both simple and complex oligosaccharide moieties.

The life cycle of this virus has been examined with the use of baby hamster kidney (BHK) cells. It first binds to the plasma membrane before being internalized by coated pits and is ultimately delivered to the lysosomal compartment. Here the low pH causes a rapid and specific fusion of the viral and lysosomal membranes so that the nucleocapsid is expelled into the cell cytoplasm (Helenius et al. 1980). This is the infection stage and it is here that the RNA molecule unravels and transcription and translation occur. The three glycoproteins E_1, E_2 and E_3 are synthesized on membrane-bound ribosomes (Garoff et al. 1978) and then transported through the Golgi apparatus to the cell surface (Green et al. 1981). There they are recruited by underlying nucleocapsids to form new intact virus particles, which then bud off from the plasma membrane.

An advantage of using a viral system is that the virus turns off host cell protein synthesis and creates a system that is devoted to conveying large numbers of just a few types of membrane protein from the endoplasmic reticulum to the cell surface. It is known that approximately 30 000 spikes are synthesized per minute in an infected BHK cell and since it takes at least 30 min for these to move to the surface, there may be as many as a million spikes moving through the cell at any one time.

To examine the route taken by the viral glycoproteins, Warren and co-workers used immunofluorescence techniques. They treated fixed, permeabilized cells with antibodies to the viral glycoproteins and visualized the first antibody with a second antibody conjugated to either rhodamine or fluorescein. To locate the viral proteins in a particular organelle, they carried out double-labelling experiments with fluorescent antibodies specific to the endoplasmic reticulum or Golgi apparatus (Louvard et al. 1982) in conjunction with the fluorescent anti-viral antibodies. In this way they were able to follow the viral proteins from one compartment to another as defined by antigenic determinants rather than morphological criteria.

To quantitate viral protein movement, Warren and co-workers used the e.m. technique of thin (0.1 μm) frozen sections, labelled with anti-viral antibody and visualized with protein A conjugated to gold. This gives a static picture; to observe the movement of viral proteins the cells were treated with cycloheximide, which stops synthesis but not intracellular transport. At different times thin frozen sections were prepared and the amount of label over each of the intracellular membranes was determined. For the viral proteins the half-times for leaving the endoplasmic reticulum and Golgi apparatus were 10 and 22 min respectively (Green *et al.* 1981).

It is well established that as glycoproteins move through the Golgi stack they acquire the complex oligosaccharides, which make up 40 % of the viral glycoprotein oligosaccharide moieties. The site of this trimming and addition of sugars within the Golgi apparatus is unknown. *Ricinus communis* agglutinin I (r.c.a.) is a lectin specific for the galactose residues of complex oligosaccharides and could be used to detect these oligosaccharides as soon as they were formed within the Golgi stack. Frozen thin sections were prepared and labelled with r.c.a. followed by an anti-lectin antibody and protein A–gold. This labelled about two-thirds of the Golgi cisternae and always left one or two cisternae on the *cis* side of the Golgi stack unlabelled (Griffiths *et al.* 1982 *a*).

The labelling patterns with r.c.a. in conjunction with three cytochemical marker enzymes, glucose-6-phosphatase, thiamine pyrophosphatase and acid phosphatase allowed Warren and co-workers to delineate tentatively three distinct Golgi compartments: *cis*, medial and *trans* Golgi compartments. Each was composed of one or two cisternae. Further evidence for the existence of distinct medial cisternae came from experiments with the sodium ionophore monensin, which has been shown to block the intracellular transport of proteins without affecting their synthesis (Tartakoff & Vassalli 1978). The advantage of this ionophore in the viral system is that it causes viral proteins to accumulate in the Golgi compartment before the transport block. Owing to their viral nature these proteins bind nucleocapsids by a process that would normally only occur at the plasma membrane surface. This allowed the direct visualization under the e.m. of two distinct classes of Golgi cisternae, those that were smooth and presumably occurred topologically after the monensin block, and those that Warren calls i.c.b.ms (intracellular capsid-binding membranes), which occurred topologically before the monensin block. These i.c.b.ms all have the cytochemical characteristics attributed to the putative medial Golgi compartment. This demonstrated that even though monensin also caused swelling and distortion of the Golgi cisternae, their cytochemical integrity was mintained (Griffiths *et al.* 1982 *b*).

To investigate the precise location of Golgi associated enzyme activities within the stack, Warren and co-workers were able to exploit the fact that, owing to the presence of nucleocapsids, the i.c.b.ms were much denser in sucrose gradients than smooth cisternae. If parallel experiments in the presence or absence of monensin were carried out, the medial Golgi compartment (i.c.b.ms) could be selectively shifted down a sucrose gradient after treatment of infected cells with the ionophore. Of three activities of the Golgi apparatus – fatty acid acylation α-1,2-mannosidase and galactosyltransferase – the first was the only activity that shifted down the gradient after treatment with monensin. This suggested that fatty acid acylation was an activity of the *cis* or medial Golgi compartment. The other two were associated with the *trans* Golgi compartment, which might be expected because they were both involved in the synthesis of complex oligosaccharides, which the labelling with r.c.a. had shown was confined to *trans* cisternae (Quinn *et al.* 1982).

In conclusion Dr Warren stressed the value of raising antibodies to cell membrane organelles. These were prepared by the injection of a relatively crude membrane fraction into a rabbit. The crude antisera raised could be cleaned up by specifically adsorbing out all the undesired antibody activities onto the membrane and secretory proteins usually discarded during the preparation of the original purified membrane fraction.

These antibodies were very specific to one type of membrane organelle; indeed Warren showed evidence for an anti-Golgi antibody that recognized only a single protein of the *trans*

Golgi compartment (Louvard *et al.* 1982). Using these powerful tools they hoped to study the assembly and disassembly of complex organelles such as the Golgi apparatus during the life cycle of the cell.

The final speaker of the morning was Dr M. Lord of the University of Bradford who moved the discussion onto the subject of plant glycoproteins and 'Protein glycosylation in the castor bean endosperm'. In the castor bean system the primary metabolic event during seed germination is the breakdown of stored triglycerides, which compose 60–70 % of the total dry mass of mature seeds. These triglycerides are broken down to hexoses and serve as the energy source for the developing embryo until such time as it becomes photosynthetically competent. The metabolic pathway of this breakdown involves the conversion of fatty acids to acetyl CoA units, which then enter a glyoxylate cycle within specialized organelles called glyoxysomes. These in turn produce succinate, which enters mitochondria and is converted to oxaloacetate before entering the cytoplasm where it is ultimately converted to sucrose.

The glyoxysome is analogous in plants to the animal peroxisome (or microbody) and shares many common biochemical markers such as β-oxidation, catalase and hydroxyacid oxidase, in addition to its unique glyoxylate cycle enzymes. Its morphology is very similar to a peroxisome and it is believed to be assembled in an analogous way.

The castor bean system is appropriate for studying protein synthesis and organelle assembly for several reasons that have been exploited by Lord and his co-workers. The endosperm of the castor bean is a short-lived tissue that only lasts for about 7 days at 30 °C after water imbibition. There is no net protein synthesis and no cell division. Reserve proteins from protein bodies are broken down to their constituent amino acids from which a range of gluconeogenic enzymes, and the organelles that contain them, are synthesized. These are rapidly synthesized over 3–4 days and vanish as the tissue becomes photosynthetic. The tissue can be easily fractionated and after a differential spin to remove plastid inclusions, Lord and co-workers found that three major organelle fractions could be isolated on sucrose gradients: endoplasmic reticulum, mitochondria and glyoxysomes. The last of these could be isolated in an undamaged state and represented 20–25 % of the total particulate protein.

The problem that Lord and co-workers addressed was to determine the mechanism of assembly of the glyoxysomes in this system. If the mechanism initially proposed for microbody biogenesis occurred, then the various matrix and membrane proteins would be synthesized on membrane-bound polysomes and either co-translationally discharged across the endoplasmic reticulum membrane or inserted into it. These would then collect in a hypothetical region of the endoplasmic reticulum from which vesicles formed the microbodies by budding (de Duve & Baudhuin 1966).

However, Lord and other workers have shown that matrix proteins such as isocitrate lyase or malate synthase appear to be exclusively synthesized on free polysomes (Kindl *et al.* 1980; Zimmerman & Neupert 1980; Roberts & Lord 1981 *a*). These proteins are initially released into the cytosol and ultimately accumulate in the glyoxysome. Some support for an endoplasmic reticulum origin of at least some of the microbody membrane comes from experiments on *N*-glycosylated glycoproteins, which they find to be present in the glyoxysomal membrane (Mellor *et al.* 1980; Bergner & Tanner 1981). These are believed to be synthesized at the level of the endoplasmic reticulum and glycosylated in a co-translational manner after membrane insertion or vectorial discharge across the membrane.

As a first step to show that the glycoproteins that they had identified as glyoxysomal did not

[5]

arise from cross-contamination by other organelles, they used marker enzymes to identify membrane fractions. Choline phosphotransferase, fumarase and catalase were taken as marker enzymes for endoplasmic reticulum, mitochondria and glyoxysomes respectively. Electron microscope examination of isolated membrane fractions gave them a further measure of the homogeneity of their fractions. If they fractionated [^{35}S]methionine-labelled endosperm homogenate on sucrose gradients, remixed the isolated membranes with fresh non-radioactive carrier homogenate and recentrifuged, they found that their radioactive membranes all purified again as single peaks.

Analysis by g.l.c. of glycopeptides extracted from KCl-washed endosperm membranes indicated that a variety of sugars characteristic of N-linked glycoproteins were present (Mellor et al. 1980). Work on lipid intermediates had suggested that castor bean was similar to other systems in its ability to synthesize N-linked glycoproteins by a dolichol-mediated pathway. These synthetic pathways were restricted to the endoplasmic reticulum and could be inhibited with tunicamycin.

Obvious qualitative differences between the protein component of the three organelle fractions could be seen on 10% sodium dodecyl sulphate polyacrylamide gel electrophoresis, which again indicated that pure fractions had been obtained. After labelling with radioactive sugar precursors, clear differences in the glycoprotein complement of the organelle fractions were found. These glycoproteins were fractionated by Triton-X114, a detergent that has been shown to separate membrane proteins into those which are hydrophobic (integral) or hydrophilic (peripheral) (Bordier 1981). Perhaps more surprisingly, they found identical bands between endoplasmic reticulum and glyoxysomes and more especially between glyoxysomes and mitochondria. The bulk of these glycoproteins appeared in the hydrophobic (integral) Triton-X114 fraction as expected, and when glycosidases were used to probe the topology of the sugar moieties of these glycoproteins, they appeared to be expressed at the luminal face.

Membranes were prepared from endosperm tissue that was pulsed with radioactive N-acetyl glucosamine and chased with non-radioactive N-acetyl glcosamine for various times, and were extracted with Triton-X114 at each time point and the radioactivity measured. The endoplasmic reticulum became radioactive first and began to chase out, whereas incorporation into glyoxysomes and mitochondria increased over the time course.

From these data Lord and co-workers tentatively concluded that the glycoproteins of the glyoxysome and the membranes in which they are inserted, could have come from endoplasmic reticulum whereas the matrix proteins appeared to enter the organelle at a later stage of development. Finally Dr Lord used the synthesis of *Ricinus communis* agglutinin (r.c.a.) as a clearcut example of a castor bean glycoprotein that is synthesized on membrane-bound polysomes and that has a site of glycosylation exclusively in the endoplasmic reticulum. It is encoded as a single large 60 kDa polypeptide that is first cleaved to remove an N-terminal 'signal sequence' before the cleavage into subunits of the ricin (toxin) and r.c.a. molecules (Roberts & Lord 1981 b).

Professor Northcote then opened up the three papers for general discussion. A lively question period ensued in which some controversial aspects of glycoprotein synthesis and membrane biology were considered. It was hoped that some of these important questions would be raised again at the following two-day meeting.

In summary, these three lectures served as an ideal preface to the Royal Society Discussion

Meeting and, on behalf of all the participants, Professor Northcote expressed gratitude to Dr R. Porter and the Ciba Foundation for their generous hospitality.

REFERENCES

Bergner, U. & Tanner, W. 1981 Occurrence of several glycoproteins in glyoxysomal membranes of castor beans. *FEBS Lett.* **131**, 68–72.

Bordier, C. 1981 Phase separation of integral membrane proteins in Triton X-114 solution. *J. biol. Chem.* **256**, 1604–1607.

de Duve, C. & Baudhuin, P. 1966 Peroxisomes (Microbodies and related particles). *Physiol. Rev.* **46**, 323–357.

Garoff, H., Simons, K. & Dobberstein, B. 1978 Assembly of the Semliki Forest virus membrane glycoproteins in the membrane of the endoplasmic reticulum *in vitro*. *J. molec. Biol.* **124**, 587–600.

Green, J., Griffiths, G., Louvard, D., Quinn, P. & Warren, G. 1981 Passage of viral membrane proteins through the Golgi complex. *J. molec. Biol.* **152**, 663–698.

Griffiths, G., Brand, R., Burke, B., Louvard, D. & Warren, G. 1982*a* Viral membrane proteins acquire galactose in *trans* Golgi cisternae during intracellular transport. *J. Cell Biol.* (In the press.)

Griffiths, G., Quinn, P. & Warren, G. 1982*b* Dissection of the Golgi complex. I. Monensin inhibits the transport of viral membrane proteins from *cis* to *trans* Golgi cisternae. *J. Cell Biol.* (Submitted.)

Helenius, A., Kartenbeck, J., Simons, K. & Fries, E. 1980 On the entry of Semliki Forest virus into BHK-21 cells. *J. Cell Biol.* **84**, 404–420.

Kindl, H., Köller, W. & Frevert, J. 1980 Cytosolic precursor pools during glyoxysome biosynthesis. *Hoppe-Seyler's Z. physiol. Chem.* **361**, 465–467.

Koch, G. L. E. & Smith, M. J. 1978 An association between actin and the major histocompatibility antigen H-2. *Nature, Lond.* **273**, 274–278.

Louvard, D., Reggio, H. & Warren, G. 1982 Antibodies to the Golgi complex and the rough endoplasmic reticulum. *J. Cell. Biol.* **92**, 92–107.

Mellor, R. B., Krusius, T. & Lord, J. M. 1980 Analysis of glycoconjugate saccharides in organelles isolated from castor bean endosperm. *Pl. Physiol.* **65**, 1073–1075.

Quinn, P., Griffiths, G. & Warren, G. 1982 Dissection of the Golgi complex. II. Separation of Golgi functions in virally-infected cells using monensin. *J. Cell Biol.* (Submitted.)

Roberts, L. M. & Lord, J. M. 1981*a* Synthesis and post-translational segregation of glyoxysomal isocitrate lyase from castor bean endosperm. *Eur. J. Biochem.* **119**, 43–49.

Roberts, L. M. & Lord, J. M. 1981*b* The synthesis of *Ricinus communis* agglutinin. Cotranslational and post-translational modification of agglutinin polypeptides. *Eur. J. Biochem.* **119**, 31–41.

Tartakoff, A. & Vassalli, P. 1978 Plasma cell immunoglobulin secretion. Arrest is accompanied by alterations in the Golgi complex. *J. exp. Med.* **146**, 1332–1345.

Zimmerman, R. & Neupert, W. 1980 Biogenesis of glyoxysomes. Synthesis and intracellular transfer of isocitrate lyase. *Eur. J. Biochem.* **112**, 225–233.

Phil. Trans. R. Soc. Lond. B **300**, 117–127 (1982)
Printed in Great Britain

Studies on the synthesis and processing of the asparagine-linked carbohydrate units of glycoproteins

By R. G. Spiro and Mary Jane Spiro

Departments of Biological Chemistry and Medicine,
Harvard Medical School,
and the Elliott P. Joslin Research Laboratory,
One Joslin Place, Boston, Massachusetts 02215, U.S.A.

It has become apparent in recent years from the work of a number of laboratories that the N-glycosylation of both membrane and secretory glycoproteins is effected by the transfer *en bloc* to nascent polypeptides of a glucose-containing oligosaccharide ($Glc_3Man_9GlcNAc_2$) from a dolichyl pyrophosphoryl carrier; this is followed by a series of modifying reactions to yield the mature polymannose and complex asparagine-linked carbohydrate units.

The enzymic steps involved in the assembly of the precursor oligosaccharide, its transfer to protein and its subsequent processing represent potential sites for the regulation of glycoprotein synthesis. Studies performed in our laboratory have dealt primarily with thyroid slices and particulate enzymes with special regard to the role of glucose in these events. Thyroglobulin, the major secretory glycoprotein of this tissue, has well defined complex and polymannose saccharide units, and indeed the most complete form of the latter ($Man_9GlcNAc_2$) has the same structure as the lipid-linked oligosaccharide without the glucose.

Our studies indicate that effective N-glycosylation requires a complete glucose chain (Glc_3) and that the glucose sequence is assembled from dolichol-P-glucose in a stepwise manner through the concerted action of at least two transferases in a fashion complementary to the subsequent excision of this sugar by glucosidases. Pulse–chase studies indicate that, after the transfer to protein, the removal of all three glucose residues as well as of the first mannose takes place in the endoplasmic reticulum and three additional mannoses are excised in the Golgi complex, because in the presence of an inhibitor of intracellular transport, carbonyl cyanide *m*-chlorophenylhydrazone (CCCP), there is a pronounced accumulation of protein-linked $Man_8GlcNAc_2$.

Studies with metabolic inhibitors (CCCP, antimycin, N_2) indicate that, under conditions of energy depletion, glucosylation of oligosaccharide-lipid is selectively impaired, resulting in an accumulation, as measured chemically or metabolically, of high-mannose-containing ($Man_9GlcNAc_2$ and $Man_8GlcNAc_2$) lipid-linked saccharides. Further evidence that the glucosylation reaction is very sensitive to the metabolic state is suggested by the observation that tissues not rapidly frozen after removal from the animal show a similar depletion of the glucose-containing oligosaccharide lipids.

Another important aspect for the regulation of N-glycosylation of proteins is the availability of dolichyl phosphate for the formation of the lipid-linked mono- and oligosaccharides. Our studies with puromycin suggest that there is a limited supply of the lipid carrier, because in the presence of this inhibitor there is no accumulation of any of the oligosaccharide-lipid species.

Introduction

The biosynthesis of the asparagine-linked carbohydrate units of glycoproteins has been a subject of active investigation in recent years and constitutes one of the most interesting chapters in the field of conjugated saccharides. The N-linked carbohydrate units were initially recognized through the work on ovalbumin by Neuberger and his colleagues (Johansen *et al.* 1961), and our work on thyroglobulin first indicated that two quite distinct types of asparagine-linked units occur (Spiro 1965). These two forms, which were originally designated unit A (simple) and unit B (complex) and are now frequently referred to as polymannose and complex (figure 1), have been found to be widely distributed among glycoproteins of diverse function

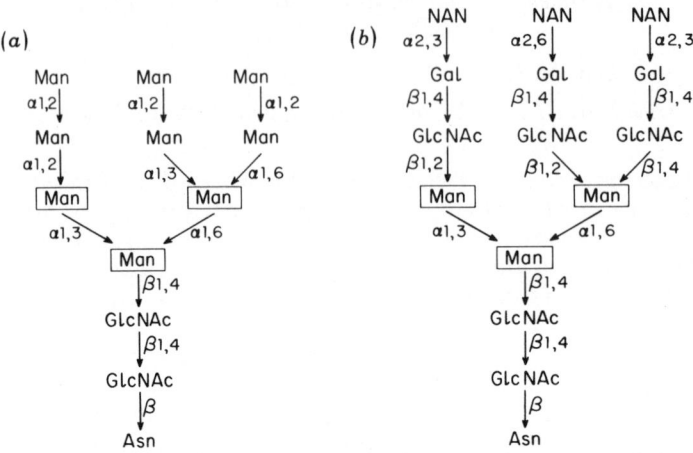

FIGURE 1. Structure of two major forms of asparagine-linked carbohydrate units of glycoproteins. The examples shown are (*a*) the polymannose unit (unit A) of calf thyroglobulin and (*b*) the complex unit from fetuin. The three internal mannose residues common to both types of saccharide unit are boxed.

and origin (Spiro 1973). A comparison of the two units indicates an interesting structural identity of their pentasaccharide ($Man_3GlcNAc_2$) core portions (figure 1). Complex carbohydrate units vary in the number and completion of the sialyl-*N*-acetyllactosamine chains and the presence or absence of fucose substituents; even the simplest variant, the pentasaccharide itself, has been observed (Dunn & Spiro 1967). More recently complex units containing repeating *N*-acetyllactosamine sequences (Järnefelt *et al.* 1978) and chains with sulphated sugars (Bedi *et al.* 1982) have been reported. Although the variations of the complex saccharide units are believed to be a function of differences in glycosylation of a common core, the several forms of the polymannose unit observed ($Man_9GlcNAc_2$ to $Man_5GlcNAc_2$) probably represent degrees of processing. The phosphorylated polymannose units of lysosomal enzymes, however, are the product of a specific transfer reaction (Reitman & Kornfeld 1981), and the 'hybrid' units, such as are seen in ovalbumin, appear to be the result of the addition of peripheral sugars characteristic of complex carbohydrate units to an incompletely processed polymannose oligosaccharide (Harpaz & Schachter 1980).

   The work of a number of laboratories has now provided an outline of the elaborate pathways (figure 2) that lead to the synthesis of the asparagine-linked saccharide units of glycoproteins (For a recent review see Hubbard & Ivatt (1981).) The early investigations of Leloir and his coworkers (Behrens & Leloir 1970), which demonstrated the presence of dolichol-linked saccharide intermediates in eukaryotic cells, paved the way for an understanding of the mechanism of assembly of this type of unit. Subsequent studies indicated that *N*-glycosylation is accomplished through the transfer *en bloc* of an oligosaccharide from a dolichyl pyrophosphate carrier to nascent polypeptide and that both dolichyl phosphate sugars and nucleoside diphosphate sugars participate in the stepwise assembly of the lipid-linked saccharide unit (figure 2).

STRUCTURE AND FUNCTION OF GLUCOSE-CONTAINING OLIGOSACCHARIDE-LIPIDS

Perhaps the most fascinating aspect of asparagine-linked carbohydrate biosynthesis is the occurrence on the transferred oligosaccharide of glucose that is subsequently removed and does not appear on the mature carbohydrate unit. The first clear indication that glucose was part of a dolichol-linked oligosaccharide containing the polymannose-di-N-acetylchitobiose

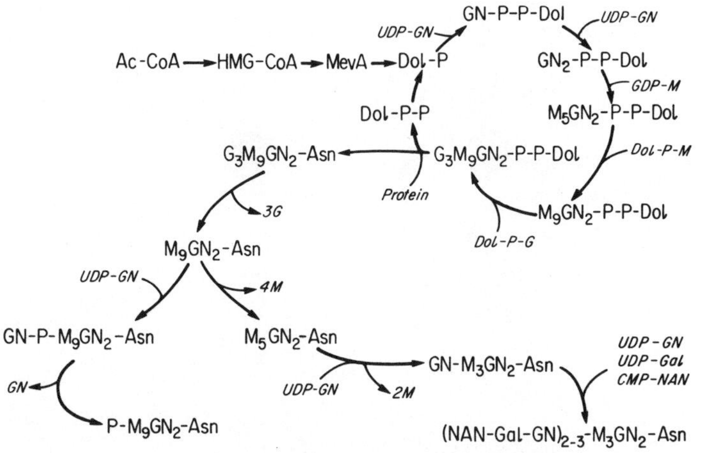

FIGURE 2. Scheme of the assembly and processing of asparagine-linked carbohydrate units of glycoproteins. The pathways leading to the synthesis of the complex, polymannose and phosphorylated polymannose units are shown. Non-standard abbreviations used are: G, glucose; M, mannose; GN, N-acetylglucosamine; NAN, N-acetylneuraminic acid; Dol, dolichol; MevA, mevalonic acid; HMG-CoA, hydromethylglutaryl coenzyme A.

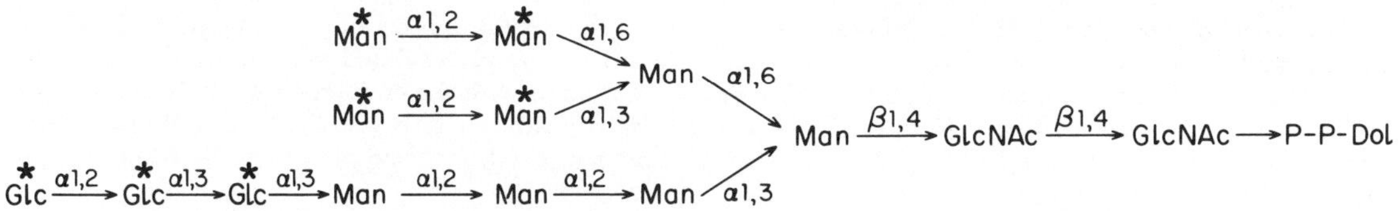

FIGURE 3. Structure of the dolichol-linked oligosaccharide that is believed to be the physiological N-glycosylating agent. The sugars marked with asterisks are derived by transfer from dolichol monosaccharide donors; the others are transferred from nucleoside diphosphate sugars.

unit was provided by our studies on thyroid (Spiro *et al.* 1974; M. J. Spiro *et al.* 1976). The peripheral location of this sugar in the oligosaccharide was indicated by the fact that it prevented the release by α-mannosidase of over half of the mannose residues (R. G. Spiro *et al.* 1976). The lipid-linked oligosaccharide was larger than even the most complete polymannose (unit A) variant of thyroglobulin and we postulated that glucose removal and modification by mannosidases would have to take place after transfer to protein to yield the mature N-linked carbohydrate units. A number of processing enzymes, both glucosidases and mannosidases, have now been described that do indeed carry out this remodelling function.

Studies on lipid-linked oligosaccharides from CHO cells (Li *et al.* 1978) led to a complete structural formulation (figure 3), which has proved to be consistent with observations made in a number of laboratories on eukaryotic cells ranging from yeast to higher vertebrates (Hubbard & Ivatt 1981). The structure of this precursor oligosaccharide appears to be identical to that of the most complete form ($Man_9GlcNAc_2$) of the unit A from calf thyroglobulin (Arima & Spiro 1972; Ito *et al.* 1977) with the exception that a triglucosyl sequence is attached to the polymannose chain, which is linked by an $\alpha(1 \to 3)$ bond to the most internal mannose residue. On the basis of chromium trioxide oxidation, α-anomeric configurations have been assigned to the glucose residues (R. G. Spiro *et al.* 1979). Acetolysis, which cleaves 1,6 linkages

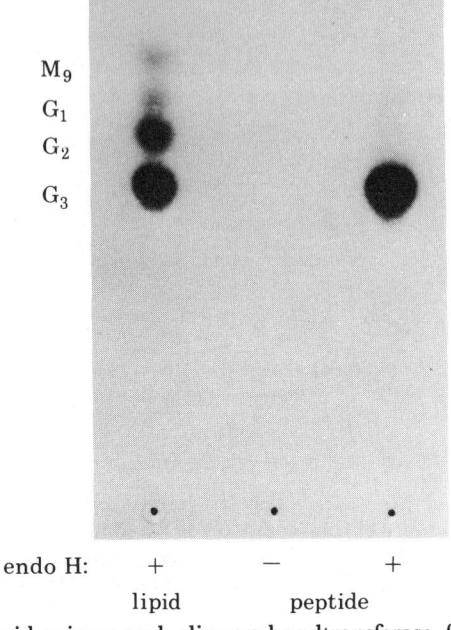

endo H: +   −   +

lipid   peptide

FIGURE 4. Specificity of thyroid microsomal oligosaccharyltransferase for fully glucosylated oligosaccharide-lipid. The enzyme was incubated with ^{14}C-labelled oligosaccharide-lipid from thyroid slices in the presence of an exogenous peptide acceptor. Thin-layer chromatographic separation on silica gel plates (1-propanol–acetic acid–water, 3:3:2) of the oligosaccharides released from the glycosylated peptide by digestion with endo-β-N-acetylglucosaminidase (endo H) is shown and compared with the endo H-treated oligosaccharides released by mild acid hydrolysis from the lipid donor. (R. J. Chalifour & R. G. Spiro, unpublished data.) The components were revealed by autoradiography. The abbreviations used are: G_3, $Glc_3Man_9GlcNAc$; G_2, $Glc_2Man_9GlcNAc$; G_1, $Glc_1Man_9GlcNAc$; M_9, $M_9GlcNAc$.

preferentially, has proved to be a valuable tool in deciphering the structure of the oligosaccharide by yielding a mannobiose and mannotriose from the two non-glucosylated chains and leaving intact the mannose chain bearing the glucose residues (R. G. Spiro *et al.* 1976). Although all three glucose residues appear to be transferred to the lipid-linked oligosaccharide from dolichol-P-glucose (Dol-P-Glc), the attachment of mannose seems to proceed from Dol-P-Man as well as directly from GDP-Man. Studies with inhibitors and cell mutants (Kang *et al.* 1978; Chapman *et al.* 1980) have shown that the five mannose residues present in the $Man_3GlcNAc_2$ core and in the chain to which the glucoses are attached are transferred from the nucleoside diphosphate sugar whereas the others are derived from Dol-P-Man (figure 3).

We have observed in studies with thyroid microsomal preparations that the glucose-containing oligosaccharide-lipid is a highly effective N-glycosylating agent for endogenous protein (M. J. Spiro *et al.* 1979). Removal of the glucose essentially abolished transfer, but α-mannosidase treatment to release peripheral mannose residues did not reduce glycosylation. Subsequent studies have indicated that the oligosaccharyltransferase requires the intact triglycosyl sequence because the oligosaccharide-lipid containing the $Glc_2Man_9GlcNAc_2$ unit was found to have minimal transfer activity (Murphy & Spiro 1981). The selectivity of the transferase can be illustrated by noting that only $Glc_3Man_9GlcNAc_2$ becomes attached to peptide acceptor when the enzyme is incubated with a mixture of oligosaccharide-lipids (figure 4). The glucose sequence represents an intriguing molecular determinant for N-glycosylation that is discarded after fulfilling its function. Because lipid-linked $Glc_3Man_9GlcNAc_2$ appears to be the physiological saccharide donor, it is evident that the enzymatic glucosylations of the oligosaccharide-lipid provide potential sites for the regulation of the attachment of carbohydrate units to asparagine residues.

The assembly of oligosaccharide-lipid

The formation of the lipid-linked oligosaccharide is initiated by the transfer of GlcNAc-1-P to Dol-P and is believed to proceed then by an ordered stepwise addition of mannose and glucose residues (figure 2). Various intermediates have been observed (Chapman et al. 1979) but little is known about the individual enyzmes involved in this biosynthetic process. Because the complete oligosaccharide consists of 14 sugar residues, that many distinct transferases could potentially be involved.

Our laboratory has recently undertaken a study of the enzymes responsible for the assembly of the glucose sequence, and our findings so far indicate the involvement of Dol-P-Glc as the sole glycosyl donor and the concerted action of at least two enzymes, with the transferase that adds the terminal $(1 \rightarrow 2)$-linked glucose being distinct from the enzyme(s) that attach the two internal $(1 \rightarrow 3)$-linked residues (Murphy & Spiro 1981). When thyroid microsomes are incubated with radiolabelled Dol-P-Glc there is a rapid transfer of glucose to endogenous oligosaccharide-lipid acceptor, with $Glc_3Man_9GlcNAc_2$ being the major saccharide product of all times (figure 5). Inclusion of sodium chloride in the incubation results in a pronounced decrease in $Glc_3Man_9GlcNAc_2$ formation with a proportional molar increase in the less completely glucosylated lipid-linked oligosaccharides, so that $Glc_2Man_9GlcNAc_2$ becomes the major component (figure 6). Methylation of the $Glc_3Man_9GlcNAc_2$ product indicated extensive radiolabelling of internal glucose residues, which was consistent with the biosynthesis de novo of the glucose chain beginning with an endogenous glucose-free, lipid-linked acceptor (Murphy & Spiro 1981).

The nature of endogenous oligosaccharide-lipids and the effect of energy state on their biosynthesis

To characterize the endogenous oligosaccharide-lipid present in thyroid we have chemically radiolabelled the oligosaccharides by $[^3H]NaBH_4$ reduction after their release from the lipid by mild acid hydrolysis. The separation of the reduced oligosaccharides was accomplished by thin-layer chromatography under conditions previously described (Murphy & Spiro 1981) and the resolution of the various species was enhanced by performing the reduction after a digestion of the oligosaccharides with endo-β-N-acetylglucosaminidase (endo H). Chromatography of the lipid-derived oligosaccharides from unincubated thyroid brought to the laboratory on ice revealed a substantial amount of unglucosylated material, which was primarily present as $Man_9GlcNac_2$ (figure 7). However, after incubation in an oxygenated medium the oligosaccharides became depleted in $Man_9GlcNAc_2$ so that $Glc_3Man_9GlcNAc_2$ became the predominant species (figure 7), although on a molar basis the total oligosaccharide-lipid pool remained unchanged (about 2 nmol g^{-1} tissue). The molar distribution of lipid-linked oligosaccharides in thyroid microsomes was similar to that seen in the total thyroid slices (table 1); the substantial amount of $Man_9GlcNAc_2$ probably represents the unglucosylated endogenous oligosaccharide-lipid which serves as acceptor for the Dol-P-Glc:oligosaccharide-lipid glucosyltransferase.

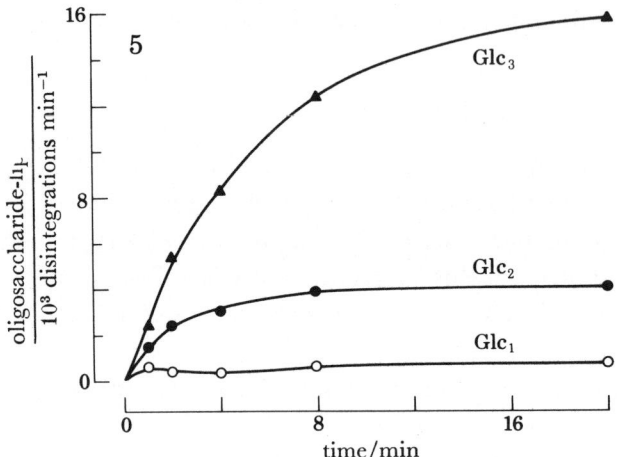
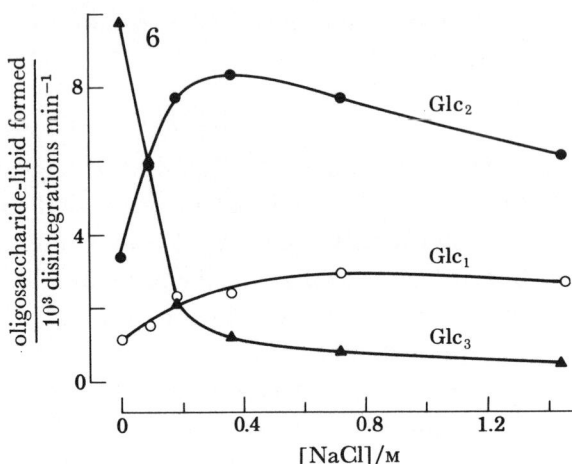

FIGURE 5. Transfer of glucose from Dol-P-[^{14}C]glucose to endogenous oligosaccharide-lipid acceptor by thyroid microsomal enzyme as a function of time. The lipid-linked oligosaccharides of the product were released by mild acid hydrolysis and separated by paper chromatography. (From Murphy & Spiro (1981).) The abbreviations are as follows: Glc_3, $Glc_3Man_9GlcNAc_2$; Glc_2, $Glc_2Man_9GlcNAc_2$; Glc_1, $Glc_1Man_9GlcNAc_2$.

FIGURE 6. Effect of sodium chloride concentration on the nature of the lipid-linked oligosaccharides formed by thyroid microsomal glucosyltransferases. The transfer of glucose from Dol-P-[^{14}C]glucose to endogenous oligosaccharide-lipid acceptor was measured. The oligosaccharides were released by mild acid hydrolysis and separated by paper chromatography. (From Murphy & Spiro (1981).) Abbreviations are the same as in figure 5.

Rapidly frozen tissue, unlike tissue kept on ice, retains its lipid-linked oligosaccharides primarily in the fully glucosylated form ($Glc_3Man_9GlcNAc_2$) and has a similar oligosaccharide pattern to that of incubated slices (table 1). Studies that we have carried out with carbonyl cyanide m-chlorophenylhydrazone (CCCP), an uncoupler of oxidative phosphorylation, suggest that glucosylation of oligosaccharide-lipid is extremely sensitive to the energy state of the tissue and that this may account for the reversible pile-up of $Man_9GlcNAc_2$ in poorly oxygenated tissue. In the presence of CCCP there is an inhibition of glucosylation with an accumulation of $Man_9GlcNAc_2$ and to a lesser extent $Man_8GlcNAc_2$ (figure 7); a similar situation is observed when slices are incubated under anaerobic conditions (table 1) or in the presence of antimycin (data not shown). In all these cases the total oligosaccharide-lipid pool remains unchanged but there is a pronounced depletion of the glucosylated components (table 1), which suggests that, while $Glc_3Man_9GlcNAc_2$-lipid is still available, transfer to protein remains relatively unaffected in these energy-deprived states. A calculation of the average number of moles of glucose per mole of oligosaccharide-lipid under various conditions is quite revealing (table 1). The unincubated tissue contains oligosaccharide-lipids with an average of 1.3 glucose residues, which is similar to the value previously observed on the basis of sugar analyses performed on

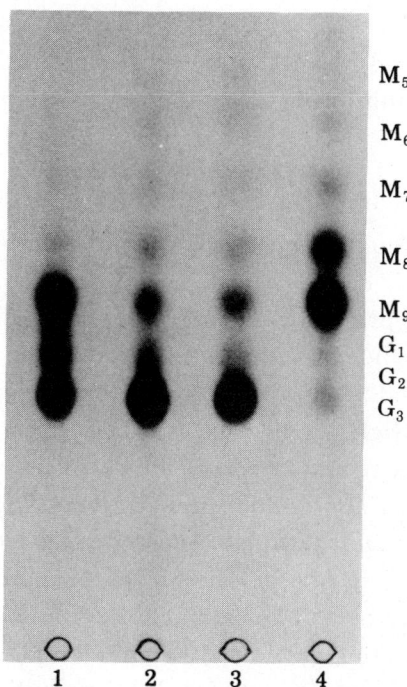

FIGURE 7. Effect of energy status on lipid-linked oligosaccharides of thyroid. Thyroids were brought to the laboratory on ice and sliced. Incubations were carried out in an oxygenated medium containing unlabelled pyruvate (5 mM) for 60 min (sample 2) and 150 min (sample 3). Sample 1 represents unincubated slices whereas the slices in sample 4 were incubated for 90 min in the presence of CCCP (100 μM) after a pre-incubation of 60 min. The lipid-linked oligosaccharides were separated by thin-layer chromatography on silica gel plates (1-propanol–acetic acid–water, 3:3:2) after release by mild acid hydrolysis, endo H digestion and reduction with [³H]NaBH₄. The components were revealed by autoradiography. The abbreviations G_3, G_2, G_1 and M_9 are defined in figure 4; additional abbreviations are: M_8, $Man_8GlcNAc$; M_7, $Man_7GlcNAc$; M_6, $Man_6GlcNAc$; M_5, $Man_5GlcNAc$.

TABLE 1. NATURE OF LIPID-LINKED OLIGOSACCHARIDES OF THYROID
IN VARIOUS CONDITIONS

(Oligosaccharide-lipids were measured by a chemical radiolabelling method. In this procedure oligosaccharides were released by mild acid hydrolysis, digested with endo H, reduced with [³H]NaBH₄ and then separated by thin-layer chromatography (see figure 7) before scintillation counting.)

	oligosaccharide-lipid (mol %)†					
condition	G_3	G_2	G_1	M_9	M_8	Glc/oligo‡
frozen tissue	59	21	7	7	6	2.3
slices, unincubated	30	16	11	33	10	1.3
slices, incubated§	59	11	7	15	8	2.1
slices, incubated + CCCP	13	6	8	48	25	0.6
slices, incubated + N_2	13	5	6	52	24	0.6
microsomes	33	13	7	33	14	1.3

† Small amounts of M_7–M_5 that were present are not included in the calculation of molar distribution (see figure 7 for abbreviations).

‡ Expressed as average number of glucose residues per oligosaccharide.

§ When thyroid slices were metabolically radiolabelled with [³H]mannose the molar distribution of the oligosaccharide-lipids was: G_3, 63 %; G_2, 20 %; G_1, 6 %; M_9, 8 %; M_8, 3 %.

such tissue (R. G. Spiro *et al.* 1976), whereas incubated or rapidly frozen thyroid has about 2.3 glucoses per molecule, which is consistent with the pattern observed in thyroid after metabolic labelling (Murphy & Spiro 1981). Less than one glucose residue is found on average during energy depletion induced by inhibitors or oxygen deprivation (table 1).

PROCESSING OF ASPARAGINE-LINKED CARBOHYDRATE UNITS

It has become apparent through the work of a number of investigators that after the transfer of the oligosaccharide from lipid carrier to protein there is a removal of all three glucose residues, and this is followed by further processing through the action of α-mannosidases to yield variants of the polymannose unit and the core portion of the complex carbohydrate unit. Studies from our laboratory have observed the processing of the carbohydrate of thyroglobulin from $Glc_3Man_9GlcNAc_2$ to $Man_5GlcNAc_2$ in thyroid slices (Godelaine *et al.* 1981) and enzymes that carry out these trimming reactions can be detected with the help of radiolabelled oligosaccharides or glycopeptides in microsomal preparations. The processing of the thyroglobulin carbohydrate units is, however, not complete as can be seen from an examination by thin-layer chromatography of the oligosaccharides released from the mature protein after endo H digestion and $[^3H]NaBH_4$ reduction (figure 8). The heterogeneity of the polymannose (unit A), previously observed from an examination of glycopeptides fractionated by ion-exchange chromatography (Arima *et al.* 1972), is clearly evident and a molar distribution of the various oligosaccharide species can readily be calculated from the radiolabelling pattern (figure 8). A substantial portion of the unit A is untrimmed ($Man_9GlcNAc_2$) or minimally digested by mannosidase, suggesting that steric factors or the rapidity of movement of the protein through the cellular compartments or both, may determine the final structure of these carbohydrate units.

Although the excision of the glucose from the transferred saccharide unit is believed to take place in the endoplasmic reticulum (Grinna & Robbins 1979) a number of studies have indicated that mannose removal by two distinct mannosidases is carried out in the Golgi apparatus (Tabas & Kornfeld 1979; Tulsiani *et al.* 1982). Recent pulse–chase studies carried out in our laboratory to follow the processing of the thyroglobulin carbohydrate units (Godelaine *et al.* 1981) have, however, suggested that the first mannose residue may, along with the three glucose residues, be released in the endoplasmic reticulum closer to the site of the initial carbohydrate attachment. When the chase was carried out in the presence of CCCP, which is known to block the transfer of proteins from the rough endoplasmic reticulum to the Golgi complex (Tartakoff & Vassalli 1977), there was a pronounced accumulation of thyroglobulin-bound $Man_8GlcNAc_2$ rather than $Man_9GlcNAc_2$ as might have been expected if all mannosidases were located in the Golgi (figure 9). The altered distribution of radioactivity in the endo H-releasable carbohydrate units in the presence of the CCCP is shown in figure 10.

THE REGULATION OF ASPARAGINE-LINKED CARBOHYDRATE UNIT SYNTHESIS

From the scheme depicting oligosaccharide-lipid assembly (figure 2) it is apparent that there are a number of potential regulatory points, and perhaps one of the most important is availability of the dolichyl phosphate carrier. Indeed it has been shown that dolichyl phosphate can be a limiting factor in asparagine-linked carbohydrate synthesis (Carson *et al.* 1981), and inhibition of hydroxymethylglutaryl coenzyme A reductase can block lipid-oligosaccharide synthesis (Mills & Adamany 1978). Some time ago we observed that in the presence of puromycin transfer of carbohydrate to protein from oligosaccharide-lipid was blocked and that an inhibition of sugar incorporation into lipid-saccharide donor also occurred (M. J. Spiro et al. 1976). While the first observation was consistent with the known action of puromycin, the latter finding suggested to us that the available dolichyl phosphate in thyroid was limited. We have now extended these studies by comparing the levels of the individual prelabelled

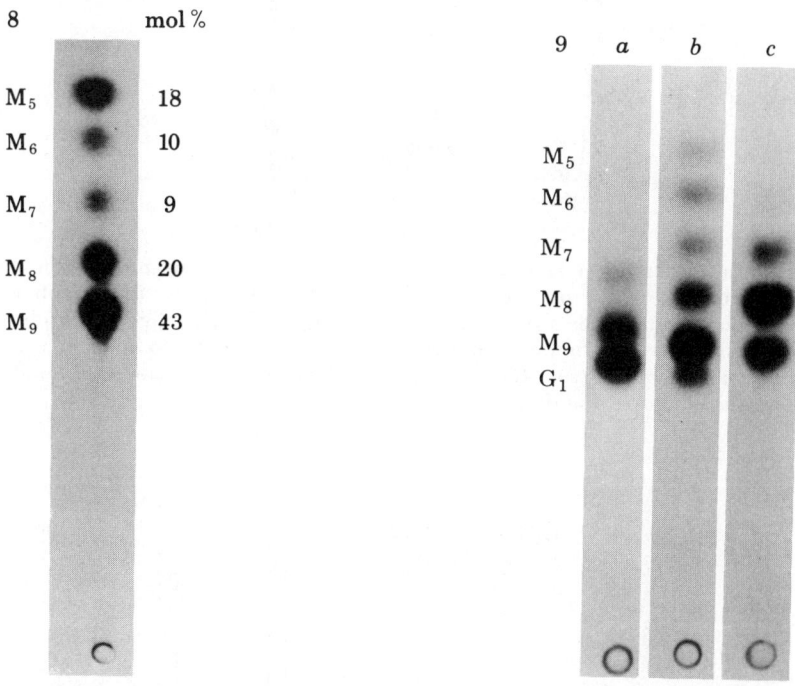

FIGURE 8. Nature and molar distribution of polymannose units of calf thyroglobulin as determined by thin-layer chromatography. The glycopeptide fraction obtained after pronase digestion of calf thyroglobulin was treated with endo H, and the released oligosaccharides were then reduced with [^{3}H]NaBH$_4$. Chromatography of the reduced, radiolabelled oligosaccharides was carried out on silica gel plates in 1-propanol–acetic acid–water, 3:3:2. The components were revealed by autoradiography and their radioactivity determined by scintillation counting after removal from the plate. The abbreviations are the same as in figures 4 and 7.

FIGURE 9. Nature of the radiolabelled oligosaccharides released by endo H treatment of thyroglobulin isolated from thyroid membranes, after a 60 min pulse with [^{14}C]glucose (a) and 180 min chase (b, c) of thyroid slices. The chase incubations were carried out in the absence (b) and presence (c) of CCCP (80 μM). Separation of the oligosaccharides was carried out by thin-layer chromatography on silica gel plates (1-propanol–acetic acid–water, 3:3:2) and the components were revealed by autoradiography. (Data taken from Godelaine et al. (1981).) The abbreviations are the same as in figures 4 and 7.

lipid-linked oligosaccharides after a chase in the presence or absence of puromycin (figure 11). Although puromycin inhibited transfer of Glc$_3$Man$_9$GlcNAc$_2$ to protein, no significant increase in this oligosaccharide-lipid was observed (figure 11). Chemical analyses of the lipid-linked oligosaccharides by the [^{3}H]NaBH$_4$ procedure gave similar results. Moreover, because there was no accumulation of smaller species of lipid-linked oligosaccharides, as determined by thin-layer chromatography, a limitation of the dolichyl phosphate available for glycosylation can again be inferred, although the possibility of feedback inhibition in the initial sugar attachment must also be considered. Dolichyl phosphate suitable for participation in lipid-oligosaccharide

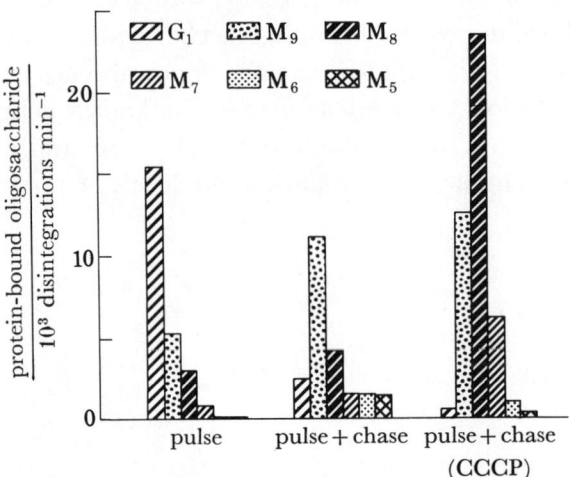

FIGURE 10. Distribution of radioactivity in oligosaccharides released by endo H from thyroglobulin isolated from membranes after incubations of thyroid slices with [14C]glucose. The pulse incubation was 60 min; the chase (180 min) was carried out in the absence and presence of CCCP. The oligosaccharides were separated by thin-layer chromatography (see figure 9) and the radioactivity in each component was determined by scintillation counting after removal from the plate. (Data taken from Godelaine et al. (1981).) The abbreviations are the same as in figures 4 and 7.

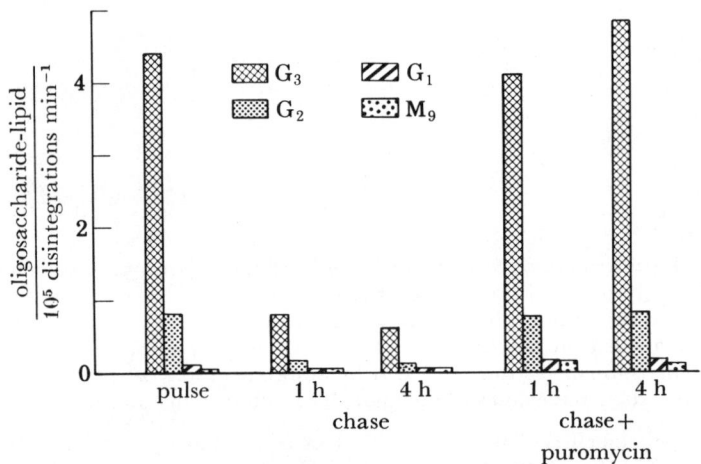

FIGURE 11. Effect of puromycin on the oligosaccharide-lipids of thyroid slices. Incubations with [14C]glucose were carried out under conditions previously described (M. J. Spiro et al. 1976). After a 1 h pulse the chase incubations were performed in the absence and presence of puromycin (0.6 mM). The lipid-linked oligosaccharides were released by mild acid hydrolysis and separated by thin-layer chromatography after endo H digestion. The radioactivity of each oligosaccharide was determined by scintillation counting after removal from the chromatographic plate. The abbreviations are the same as in figure 4.

synthesis could be generated through *de novo* synthesis, recycling or phosphorylation of the free dolichol present in tissues.

It is evident that the addition of the glucose residues to the lipid-oligosaccharide represents another important regulating factor in asparagine-linked carbohydrate unit synthesis. Because the triglucosyl sequence seems to be essential for physiological *N*-glycosylation, any interference with its assembly, as in our energy-deprived thyroid slices, would be reflected in an inhibition of saccharide unit attachment.

The processing steps (figure 2) afford further opportunities for regulation: some carbohydrate units are not reduced in size beyond $Man_9GlcNAc_2$ whereas others are trimmed further and after reaching the level of $Man_5GlcNAc_2$ they may give rise to complex oligosaccharides. Furthermore the asparagine-linked carbohydrate of lysosomal enzymes undergoes a distinctive modification involving the phosphorylation of mannose residues. The amino acid sequence of the glycosylated protein and its speed of migration to the cell surface may play a role in determining the nature and extent of the processing reactions.

Finally it must be appreciated that the assembly and processing steps take place at several intracellular sites and require translocations across membranes and movements from one compartment to another, which provides additional possibilities for the regulation of glycoprotein synthesis.

This work was supported by grant no. AM 17477 from the National Institutes of Health and grant no. PCM 77-00376 from the National Science Foundation.

REFERENCES

Arima, T. & Spiro, R. G. 1972 *J. biol. Chem.* **247**, 1836–1848.

Arima, T., Spiro, M. J. & Spiro, R. G. 1972 *J. biol. Chem.* **247**, 1825–1835.

Bedi, G. S., French, W. C. & Bahl, O. P. 1982 *J. biol. Chem.* **257**, 4345–4355.

Behrens, N. H. & Leloir, L. F. 1970 *Proc. natn. Acad. Sci. U.S.A.* **66**, 153–159.

Carson, D. D., Earles, B. J. & Lennarz, W. J. 1981 *J. biol. Chem.* **256**, 11552–11557.

Chapman, A., Fujimoto, K. & Kornfeld, S. 1980 *J. biol. Chem.* **255**, 4441–4446.

Chapman, A., Li, E. & Kornfeld, S. 1979 *J. biol. Chem.* **254**, 10243–10249.

Dunn, J. T. & Spiro, R. G. 1967 *J. biol. Chem.* **242**, 5556–5563.

Godelaine, D., Spiro, M. J. & Spiro, R. G. 1981 *J. biol. Chem.* **256**, 10161–10168.

Grinna, L. S. & Robbins, P. W. 1979 *J. biol. Chem.* **254**, 8814–8818.

Harpaz, N. & Schachter, H. 1980 *J. biol. Chem.* **255**, 4894–4902.

Hubbard, S. C. & Ivatt, R. J. 1981 *A. Rev. Biochem.* **50**, 555–583.

Ito, S., Yamashita, K., Spiro, R. G. & Kobata, A. 1977 *J. Biochem., Tokyo* **81**, 1621–1631.

Järnefelt, J., Rush, J., Li, Y.-T. & Laine, R. A. 1978 *J. biol. Chem.* **253**, 8006–8009.

Johansen, P. G., Marshall, R. D. & Neuberger, A. 1961 *Biochem. J.* **78**, 518–527.

Kang, M. S., Spencer, J. P. & Elbein, A. D. 1978 *J. biol. Chem.* **253**, 8860–8866.

Li, E., Tabas, I. & Kornfeld, S. 1978 *J. biol. Chem.* **253**, 7762–7770.

Mills, J. T. & Adamany, A. M. 1978 *J. biol. Chem.* **253**, 5270–5273.

Murphy, L. A. & Spiro, R. G. 1981 *J. biol. Chem.* **256**, 7487–7494.

Reitman, M. L. & Kornfeld, S. 1981 *J. biol. Chem.* **256**, 4275–4281.

Spiro, M. J., Spiro, R. G. & Bhoyroo, V. D. 1976 *J. biol. Chem.* **251**, 6400–6408.

Spiro, M. J., Spiro, R. G. & Bhoyroo, V. D. 1979 *J. biol. Chem.* **254**, 7668–7674.

Spiro, R. G. 1965 *J. biol. Chem.* **240**, 1603–1610.

Spiro, R. G. 1973 *Adv. Protein Chem.* **27**, 349–467.

Spiro, R. G., Spiro, M. J. & Adamany, A. M. 1974 *Biochem. Soc. Symp.* **40**, 37–55.

Spiro, R. G., Spiro, M. J. & Bhoyroo, V. D. 1976 *J. biol. Chem.* **251**, 6409–6419.

Spiro, R. G., Spiro, M. J. & Bhoyroo, V. D. 1979 *J. biol. Chem.* **254**, 7659–7667.

Tabas, I. & Kornfeld, S. 1979 *J. biol. Chem.* **254**, 11655–11663.

Tartakoff, A. M. & Vassalli, P. 1977 *J. exp. Med.* **146**, 1332–1345.

Tulsiani, D. R. P., Hubbard, S. C., Robbins, P. W. & Touster, O. 1982 *J. biol. Chem.* **257**, 3660–3668.

Phil. Trans. R. Soc. Lond. B **300**, 129–144 (1982)
Printed in Great Britain

Topological and regulatory aspects of dolichyl phosphate mediated glycosylation of proteins

By W. J. Lennarz

The Johns Hopkins University, School of Medicine, Department of Physiological Chemistry,
725 N. Wolfe Street, Baltimore, Maryland 21205, U.S.A.

From the time of their synthesis in the rough endoplasmic reticulum until they are secreted, packaged in lysosomes, or appear as membrane components at the cell surface, the polypeptide chains of N- and O-linked glycoproteins remain associated with intracellular membranes that are components of the secretory pathway. The various co-translational and post-translational modifications of the carbohydrate moieties of glycoproteins have been shown to occur within morphologically and functionally distinct regions of this complex membrane system. However, the sugar nucleotides, which serve as precursors to the oligosaccharide moieties of these glycoproteins, are synthesized almost exclusively in the cytoplasm. These findings raise a number of questions about the mechanisms involved in the transmembrane assembly of membrane and secretory glycoproteins. In this paper these questions are reviewed and recent studies directed towards providing answers to them are summarized. In addition, information related to the possible role of dolichyl phosphate in regulating the glycosylation of proteins is presented.

Introduction

Although many of the basic steps involved in the synthesis of the oligosaccharide chains of N-linked glycoproteins are now fairly well understood, a good deal remains to be learned about the topological aspects of this assembly process and the basic mechanisms that serve to regulate it. In figure 1, a number of the key steps involved in synthesis of an N- and O-linked glycoprotein are outlined. In addition, current views of the spatial and temporal aspects of this process that will be discussed in this paper are depicted. As shown, the N- and O-glycosylation of a newly synthesized polypeptide chain is a topologically asymmetric process. The process is initiated by the asymmetric synthesis of $(GlcNAc)_2$-PP-dolichol (chitobiosyl-lipid) with its carbohydrate chain facing the luminal side of the rough endoplasmic reticulum (r.e.r.). Subsequent elongation by addition of Man and Glc units leads to a complete oligosaccharide chain, still attached to dolichyl pyrophosphate and still facing the lumen of the r.e.r. Transfer of the oligosaccharide chain to a nascent polypeptide is believed to occur when the growing polypeptide containing a tripeptide, -Asn-X-Ser/Thr-, has transited the membrane so that the Asn site is accessible to the oligosaccharide-lipid. After transfer, the oligosaccharide chain undergoes the initial stages of processing, which involve the removal of glucose residues. After movement of this partly processed chain from the r.e.r. to the Golgi complex, further processing occurs if it is destined to become a complex-type chain. If it is destined to be a mature polymannose-type chain, relatively little additional processing occurs. For the complex chains, the removal of Man, followed by the addition of GlcNAc, Gal, Fuc and sialic acid residues results in the formation of a mature complex chain. Most of the available evidence also indicates that it is here in the Golgi complex that the assembly of *O*-linked chains occurs. As shown with this hypothetical protein containing both *N*- and *O*-linked chains, the assembly of the *O*-linked chain involves sequential addition of GalNAc, Gal and sialic acid residues.

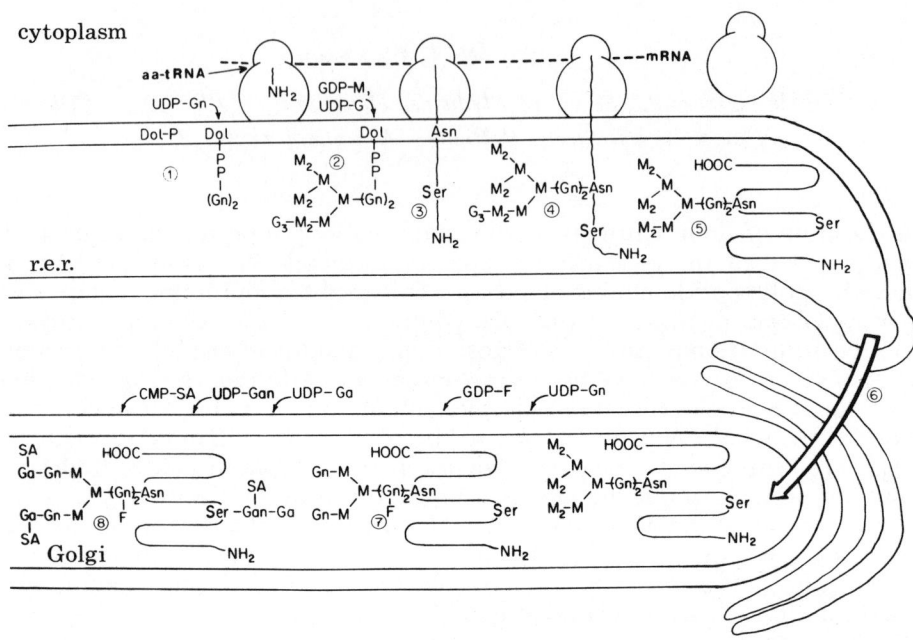

FIGURE 1. Model depicting events in the *N*- and *O*-glycosylation of a hypothetical secretory protein containing sites for both types of chains. The abbreviations are: Gn, *N*-acetylglucosamine; M, mannose; G, glucose; Ga, galactose; F, fucose; Gan, *N*-acetylgalactosamine; SA, sialic acid.

TOPOLOGICAL ORIENTATION OF SACCHARIDE-LIPIDS

A priori, there is no reason to postulate the topological orientation of the dolichol-linked saccharide shown in figure 1. An equally reasonable possibility is that assembly of these saccharide-lipids occurs at the cytoplasmic face of the r.e.r., where sugar nucleotides would be readily available. *N*-glycosylation of the growing polypeptide at this face, followed by transit of the already glycosylated protein through the membrane as translation proceeded, would result in the same end product. What, then, is the evidence that the topology is that shown in figure 1? In approaching this question, we first studied the topology of the chitobiosyl-lipid. Earlier it was observed that purified chitobiosyl-lipid was an excellent substrate for galactosyl transferase. Thus when purified chitobiosyl-lipid suspended in detergent was incubated with galactosyl transferase and UDP–Gal it was quantitatively converted from the disaccharide-lipid to a Gal-containing trisaccharide-lipid. This finding suggested the possibility that galactosyl transferase might be useful as a probe for the topological distribution of the disaccharide units of chitobiosyl-lipid in the membrane of the r.e.r. In a series of preliminary experiments, phospholipid liposomes containing small amounts of labelled chitobiosyl-lipid were prepared. These unilamellar liposomes, prepared in the two sizes shown in figure 2, were then incubated with UDP–Gal and galactosyl transferase, and the proportion of the disaccharide-lipid that was converted to trisaccharide-lipid was assessed. As shown, by using either large or small liposomes of known diameters, the extent of galactosylation was in excellent agreement with that calculated on the basis of the surface areas of the inner and outer faces of the liposome. When a detergent was added to disrupt the liposomes, all of the disaccharide units were converted to trisaccharide. Further, when this galactosylation process with liposomes was studied kinetically, with higher levels of both galactosyl transferase and UDP–Gal, it became clear that the values observed in this experiment were kinetic endpoints. That is, no further galactosylation of

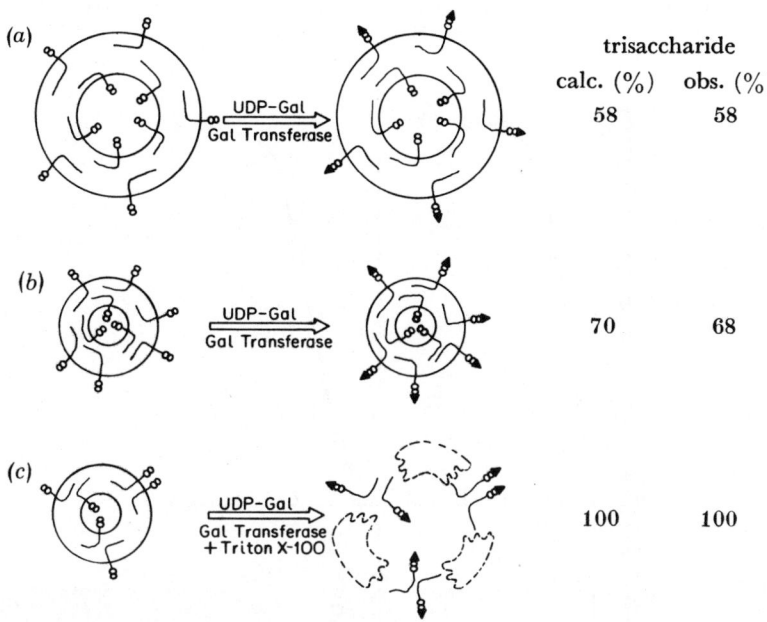

FIGURE 2. Use of galactosyl transferase as a probe for the topological distribution of chitobiosyl-lipid in phospholipid liposomes: (a) large liposomes (60 nm); (b) small liposomes (22 nm); (c) detergent-disrupted liposomes.

chitobiosyl-lipid could be observed under conditions of prolonged incubation. This strongly suggests that, at least in liposomes, the disaccharide-lipid on the inner face is unable to undergo significant flip-flop to the outer face (Hanover & Lennarz 1980).

With this information in hand, we turned to microsomes prepared from hen oviduct, and generated chitobiosyl-lipid by incubation with UDP-GlcNAc (Hanover & Lennarz 1980). Subsequently we assessed the accessibility of the preformed chitobiosyl-lipid to galactosyl transferase, using both total microsomes and purified r.e.r. As shown in figure 3 (a, b), in the absence of detergent less than 10 % of the disaccharide-lipid was accessible to galactosyl transferase, suggesting that it was present primarily at the luminal face of the membranes. However, when various amounts of detergent were added to disrupt the membranes, the disaccharide-lipid became increasingly accessible to the galactosyl transferase. At levels of detergent above 0.15 % all of the previously synthesized disaccharide units were accessible. As shown in figure 3 (c, d), in parallel experiments assessing the intactness of the membranes by measuring the latency of two known luminal enzymes, a reciprocal relation between latency and accessibility of chitobiosyl-lipid was observed. Thus in the absence of detergent, mannose-6-phosphate phosphatase and β-glucuronidase were almost completely latent. Only when the membrane was disrupted did they lose their latency and become accessible to exogenous substrates.

[23]

It was important to rule out several alternative explanations for the apparent inaccessibility of the chitobiosyl-lipid. One obvious possibility was that the disaccharide-lipid was indeed synthesized at the cytoplasmic face but was not accessible to galactosyl transferase because of the presence of proteins or ribosomes on the surface of the membranes. To test this possibility, the accessibility of chitobiosyl-lipid was measured in microsomes pretreated with proteases, such as trypsin or chymotrypsin, to remove surface proteins, or with puromycin–KCl to discharge the ribosomes. It was found that neither treatment enhanced the accessibility of the disaccharide-

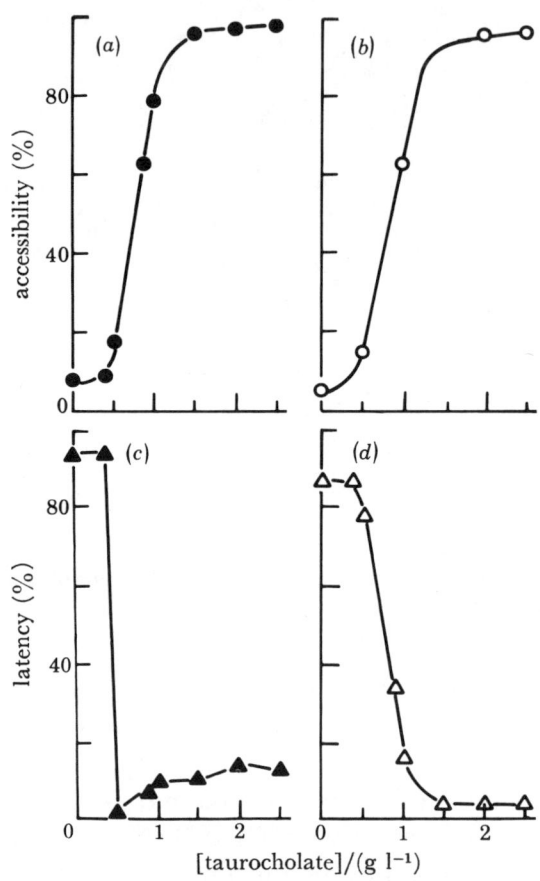

FIGURE 3. Accessibility of chitobiosyl-lipid in microsomes to galactosyl transferase in the absence or the presence of detergent: (a) total microsomes; (b) r.e.r. Also shown is the latency of two microsomal enzymes: mannose-6 phosphate phosphatase (c) and β-glucuronidase (d).

lipid to the probe. Another possibility was that the effect of detergent was selective extraction of the disaccharide-lipid from the membranes, and only after it was extracted and present in solution was it accessible to the galactosyl transferase. However, direct measurements of the release of disaccharide-lipid from the membrane as a function of the concentration of detergent excluded this possibility; no extraction of the disaccharide-lipid was observed by using the detergent concentrations employed in these experiments.

Three possible mechanisms to explain the apparently asymmetric synthesis of chitobiosyl at the luminal face of the microsomes were considered, as shown in figure 4. In A, assembly of the chitobiosyl-lipid occurs at the cytoplasmic face, but is followed by rapid, unidirectional flip-flop to the inner face. In B, assembly occurs at the luminal face but is preceded by the transport of the sugar nucleotide into the lumen. In C, the intact sugar nucleotide is not transported into the lumen. Rather, the GlcNAc and GlcNAc-1-PO_4 units that participate in synthesis of chitobiosyl-lipid are transferred from the sugar nucleotide to an enzyme complex containing

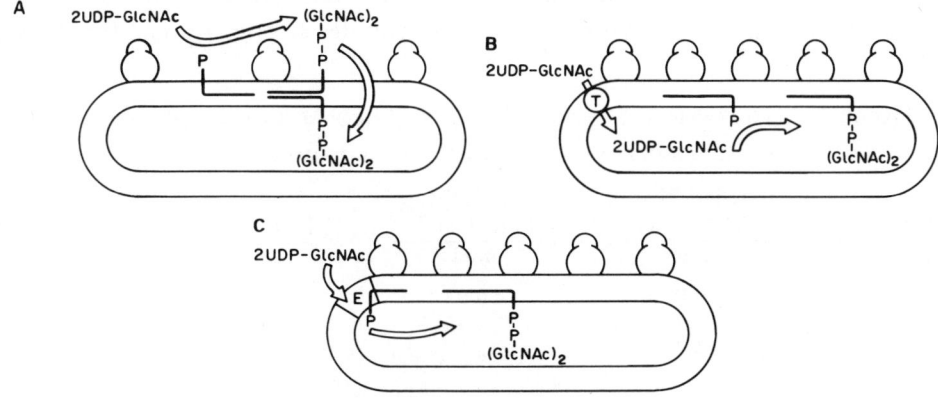

FIGURE 4. Alternative models for the asymmetric synthesis of chitobiosyl-lipid at the luminal face of microsomes.

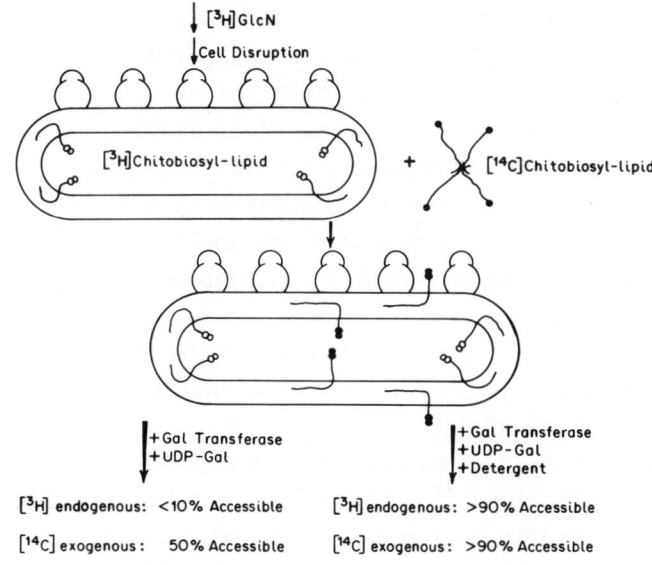

FIGURE 5. Comparison of the accessibility of chitobiosyl-lipid generated *in vivo* with the accessibility of chitobiosyl-lipid introduced into isolated microsomes.

the transferases and dolichylphosphate; subsequent transfer of the activated sugars from the putative enzyme intermediate to dolichyl phosphate results in the formation of chitobiosyl-lipid at the luminal face. In an attempt to differentiate between these possible models, and to provide further support for the conclusion that chitobiosyl-lipid is orientated luminally, the experiment outlined in figure 5 was carried out (Hanover & Lennarz 1982). Oviduct tissue slices were incubated with [³H]GlcN under conditions that lead to formation of chitobiosyl-lipid. Microsomes were then prepared from these oviduct tissue slices. These microsomes, containing

³H-labelled chitobiosyl-lipid synthesized *in vivo*, were then incubated with a highly purified preparation of ¹⁴C-labelled chitobiosyl-lipid. After centrifugation and washing of the microsomes to remove any [¹⁴C]chitobiosyl-lipid not tightly associated with the microsomes, the preparation was incubated with UDP–Gal and galactosyl transferase to study the topological distribution of both the ³H- and ¹⁴C-labelled chitobiosyl-lipid. As shown, in the absence of detergent, virtually all of the [³H]chitobiosyl-lipid synthesized *in vivo* was inaccessible to the probe until detergent was added. This observation provides strong support for the validity of

TABLE 1. UTILIZATION OF [³H]UDP–[¹⁴C]GlcNAc IN CHITOBIOSYL-LIPID
SYNTHESIS IN OVIDUCT MICROSOMES

incubation time	[¹⁴C]GlcNAc-PP-dolichol	[¹⁴C](GlcNAc)$_2$-PP-dolichol	[³H]UMP	[³H]UDP
min	pmol	pmol	pmol	pmol
0	< 0.06	< 0.06	0.25	< 0.05
30	9.7	9.45	1.6	< 0.05
60	11.7	15.9	2.0	< 0.05

the previous studies on the topology of chitobiosyl-lipid synthesized in isolated microsomes. When the topological distribution of the externally introduced [¹⁴C]chitobiosyl-lipid was examined, the results were quite different. As shown, 50% of the chitobiosyl-lipid could be galactosylated, indicating a cytoplasmic orientation. When detergent was added, all of the [¹⁴C]chitobiosyl-lipid was galactosylated. Furthermore, when galactosylation of the [¹⁴C]-chitobiosyl-lipid was studied kinetically, it was found that there was no increase in the extent of galactosylation with time, indicating that both ¹⁴C- and ³H-labelled chitobiosyl-lipid are static in the membrane, and do not undergo any significant flip-flop.

To further test the possibility that synthesis occurred on the outer face, followed by flip-flop to the inner face, we modified our original *in vitro* experiments by adding galactosyl transferase and UDP–Gal during, rather than after, synthesis of chitobiosyl-lipid in microsomes (Hanover & Lennarz 1982). Under these conditions, no significant increase in the conversion of disaccharide-lipid to trisaccharide-lipid was observed in the absence of detergent. Thus the results of both this and the previous experiment indicate that it is unlikely that synthesis occurs at the outer face as shown in model A.

In model B the sugar nucleotide is first transferred into the microsome where it reacts at the luminal face with dolichyl phosphate. However, using the techniques of Ballas & Arion (1977) to study the uptake of solutes into microsomes, we found that UDP-GlcNAc and GDP-Man behaved more like dextran, an impermeant molecule, than like glucose, a readily permeant molecule (Hanover & Lennarz 1982). In addition, it was found that under the same conditions CMP-sialic acid and UMP were readily permeable, in agreement with earlier findings by others. If the UDP-GlcNAc was taken up into the microsomes, and the transfer reactions occurred in the lumen, one would expect to find UMP and UDP, the other products of transferases, present in amounts equivalent to the chitobiosyl-lipid formed. However, as shown in table 1, this is not so: the level of the two nucleotides detected was only 10% that of the chitobiosyl-lipid formed (Hanover & Lennarz 1982).

In addition to thse findings *in vitro* arguing against model B, we carried out studies *in vivo* to measure the level of sugar nucleotides found inside the endoplasmic reticulum (Hanover &

Lennarz 1982). By using the procedure of Carey & Hirschberg (1981), it was found that little or no UDP-GlcNAc and GDP-Man could be detected in microsomes isolated from tissue slices that had synthesized the sugar nucleotides *in vivo*. Thus all of these studies are inconsistent with model B, leaving us with model C, which postulates concerted transfer of the activated sugar units, mediated by an enzyme complex interacting with dolichyl phosphate. As a direct consequence of the enzymatic transfer reaction, the observed asymmetry is introduced. Clearly, further studies will be necessary to refine and test this model.

As shown in figure 1, further elongation of the chitobiosyl-lipid leads to the assembly of the completed oligosaccharide still attached to the dolichyl pyrophosphate. Although we have not been able to develop topology probes to study the orientation of the oligosaccharide-lipid, we took advantage of the finding that under certain conditions oviduct microsomes rapidly hydrolyse a significant portion of the oligosaccharide attached to dolichyl pyrophosphate. With this information, and with the expectation that oligosaccharides this size would be impermeable to the membrane of the r.e.r., we undertook to determine where the oligosaccharide was after hydrolysis. The results of this experiment indicate that the hydrolytically released oligo-saccharide remains in the lumen and is only released from the microsomes when the membrane is disrupted with detergent (Hanover & Lennarz 1982). Thus it appears that not only the synthesis of chitobiosyl-lipid but also its elongation occurs inside the r.e.r.

TEMPORAL ASPECTS OF TRANSLATION AND GLYCOSYLATION

If the model postulated in figure 1 is correct, the polypeptide chain at the time that the Asn residue is glycosylated must be sufficiently long to transit both the ribosomal cleft and at least most of the membrane bilayer. Using oviduct tissue slices and studying ovalbumin, we under-took to determine how long the ovalbumin polypeptide chain must grow after addition of the Asn residue before this residue is glycosylated (Glabe *et al.* 1980). This question is outlined schematically in figure 6, showing the growing ovalbumin chain, the addition of Asn-293, and the presence of the carbohydrate chain attached to this residue on the mature polypeptide. Recently the nucleotide sequence of the ovalbumin gene and, independently, the amino acid sequence of the protein have been reported (Nesbit *et al.* 1981). As shown in figure 7, examina-tion of the sequence around the glycosylation site indicates that there are two tryptic cleavage sites present, which should enable one to excise a glycopeptide containing 32 amino acid residues. To ascertain that this was feasible, we subjected mature ovalbumin to trypsin diges-tion, and recovered a glycopeptide by gel filtration and Concanavalin A affinity chromato-graphy. Sequence analysis of the purified glycopeptide revealed that the first 10 residues (shown underlined) had the expected sequence. With this reference glycopeptide in hand we proceeded to study the nascent chains of ovalbumin. Oviduct tissue slices were incubated with [^{35}S]methionine and with [^{3}H]mannose, and the ovalbumin nascent chains were isolated. As outlined in figure 8, the total nascent chain mixture was subjected to trypsin digestion and the glycopeptides were isolated. If no glycopeptide smaller than 32 residues were detectable (figure 8a), one would conclude the chains must grow a *minimum* of 32 residues in length after addition of the Asn residue before the carbohydrate is added. However, if smaller glycopeptides were obtained (figure 8b), one would conclude that carbohydrate attachment occurred before the addition of 32 more residues. The isolated glycopeptides from the nascent chains were mixed with authentic glycopeptide isolated from the mature chain, and analysed by both gel filtration and sodium dodecyl sulphate polyacrylamide gel electrophoresis. As shown in figure 9, the glycopeptide from the nascent chains was identical to that from the mature chain, indicating that at least 32 residues must be added before glycosylation occurs.

In addition to asking this question, in other experiments we determined the length of the ovalbumin nascent chains that is protected in the ribosomal cleft from proteolysis by trypsin

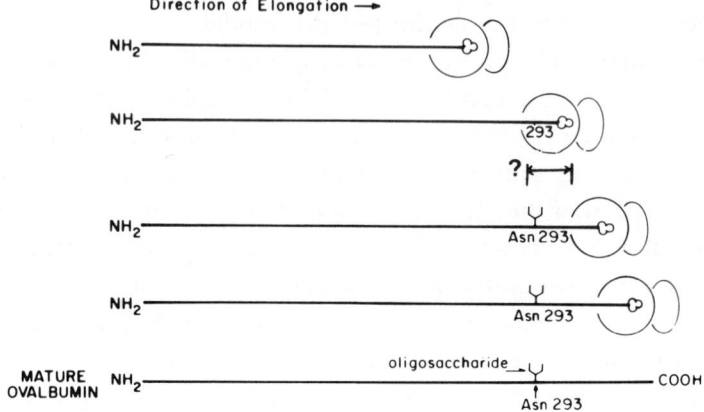

FIGURE 6. Intermediate stages in the biosynthesis of ovalbumin.

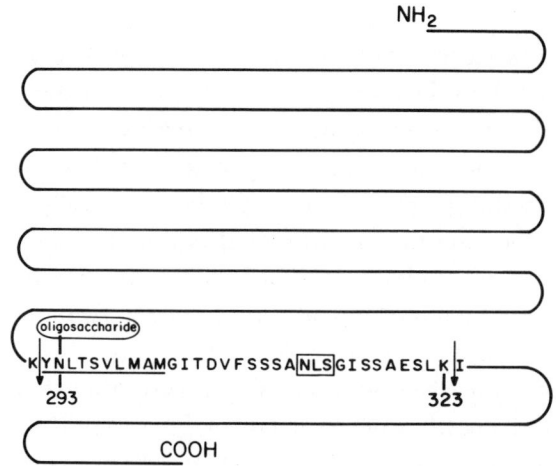

FIGURE 7. Partial primary structure of ovalbumin.

(Glabe *et al.* 1980). These experiments yielded values for a peptide approximately 20–25 residues in length, in reasonable agreement with earlier studies (Blobel & Sabatini 1970). Thus the overall minimum length of the polypeptide chain at the time of glycosylation appears to be 52 residues, i.e. 32 plus 20. This length is clearly sufficient for glycosylation of the Asn residue to occur at the luminal face of the endoplasmic reticulum, as postulated in figure 1.

O-GLYCOSYLATION

The hypothetical protein in figure 1 is depicted as having a site for *O*- as well as *N*-glycosylation. Evidence for and against the idea that *O*-glycosylation is a co-translational process has been reported (Strous 1979; Jokinen *et al.* 1979, 1981; Ruddon *et al.* 1980; Ruddon *et al.* 1981;

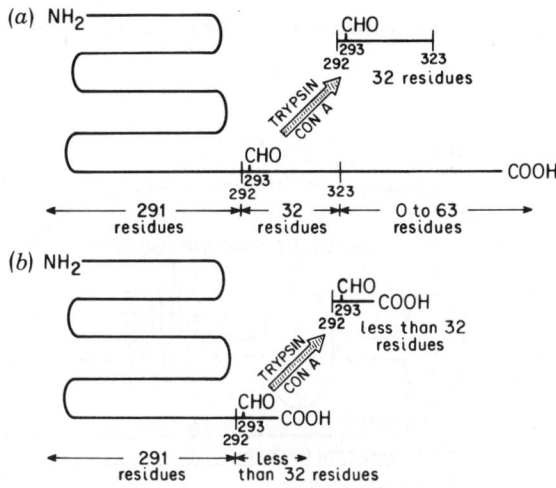

FIGURE 8. Strategy used for the isolation of glycopeptides from ovalbumin nascent chains.

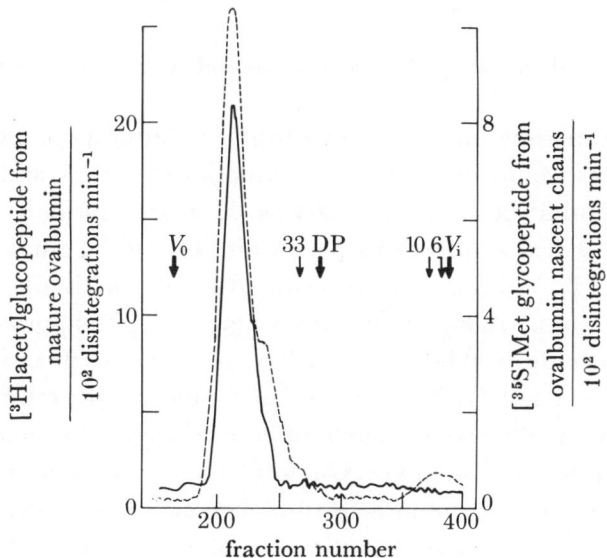

FIGURE 9. Comparison by gel filtration of the glycopeptide isolated from nascent chains (—) and from mature ovalbumin (– – –). Numbers against arrows are molecular masses in hectodaltons.

Kim *et al.* 1971; Ko & Raghupathy 1972; White & Speake 1980). Studies with oviduct membrane preparations and synthetic peptides that act as acceptors of a GalNAc unit from UDP–GalNAc indicated that the processes of *N*- and *O*-glycosylation were quite different in several respects (Hanover *et al.* 1980). First, *N*-glycosylation was stimulated by dolichyl phosphate and inhibited by tunicamycin, whereas *O*-glycosylation was unaffected by either compound.

Second, and more important in the context of whether or not *O*-glycosylation is a co-translational process, was the finding that the transferase catalysing the transfer of GalNAc to synthetic peptides was highly enriched in a smooth membrane fraction consisting of the Golgi complex and smooth endoplasmic reticulum. In contrast the enzymes involved in *N*-linked glycosylation were enriched in the r.e.r. Clearly these observations strongly suggest that unlike *N*-glycosylation, *O*-glycosylation is a post-translational process. In this context it should be noted that unlike the r.e.r., the Golgi complex appears to be permeable to nucleotides (Carey & Hirschberg 1981; Fleischer 1981).

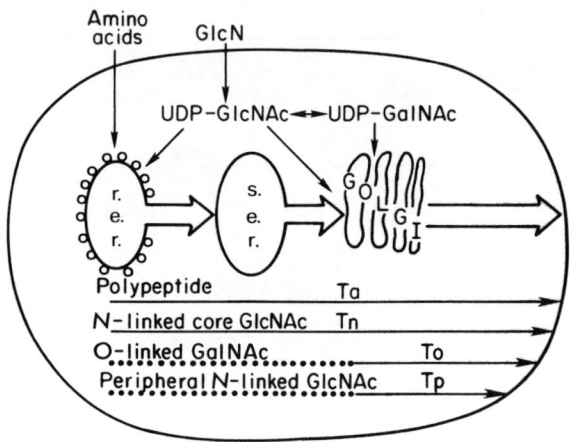

FIGURE 10. Approach for studying the synthesis and the *N*- and *O*-glycosylation of hCG in Be Wo cells.

An alternative method to determine if *N*- and *O*-glycosylation processes are temporally and spatially distinct processes would be to study the kinetics of the *N*- and *O*-glycosylation of a single polypeptide containing both types of chains. Human chorionic gonadotropin (hCG) contains both *O*- and *N*-linked chains in its β subunit. Therefore, studies with the use of the choriocarcinoma cell line Be Wo were undertaken and the synthesis and glycosylation of hCG were examined (Hanover *et al.* 1982). As shown in figure 10, the approach was to use amino acids to label the polypeptide backbone, and GlcN to label the precursors to both types of hexosamine units, namely UDP–GlcNAc and UDP–GalNAc. In preliminary experiments it was shown that GlcN was an effective precursor of both types of sugar nucleotides and that the uptake, activation and epimerization of GlcNAc to GalNAc units were rapid processes. When the kinetics of the appearance of amino acid- and hexosamine-labelled hCG in the culture medium were examined a very different profile was observed for the two precursors (figure 11*a*). No newly synthesized polypeptide chains were observed in the culture medium before 1.5 h. In contrast, hexosamine label was rapidly incorporated into the secreted hCG, well before newly synthesized, amino acid-labelled hCG was secreted. In the case of intracellular hCG (figure 11*b*), both the hexosamine and amino acid labels enter the polypeptide at similar rates.

Next we sought to compare the kinetics of incorporation of each of the three types of amino sugars in hCG, i.e. GlcNAc in the core and in peripheral positions on the *N*-linked chains, and GalNAc in the *O*-linked chains. To accomplish this the sequential degradation procedure shown in figure 12, followed by h.p.l.c. analysis of the amino sugars released after acid hydrolysis, was used. When this procedure was applied to secreted hCG isolated from the culture medium at various time intervals, the results in figure 13 were obtained. It is clear that at the earliest time interval, before any newly synthesized polypeptide chains of hCG have been secreted, virtually all of the hexosamine label is found in GalNAc units in the O-linked chain and in peripheral GlcNAc units in N-linked chains. Previous studies have shown that addition

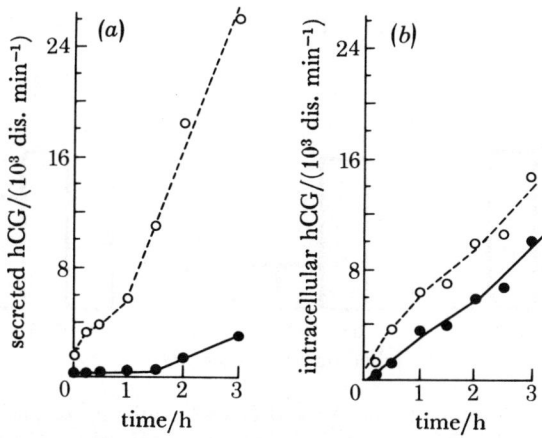

FIGURE 11. Kinetics of appearance of [^{35}S]methionine-labelled (closed circles) and hexosamine-labelled (open circles) hCG in the culture medium (a) and in the cells (b).

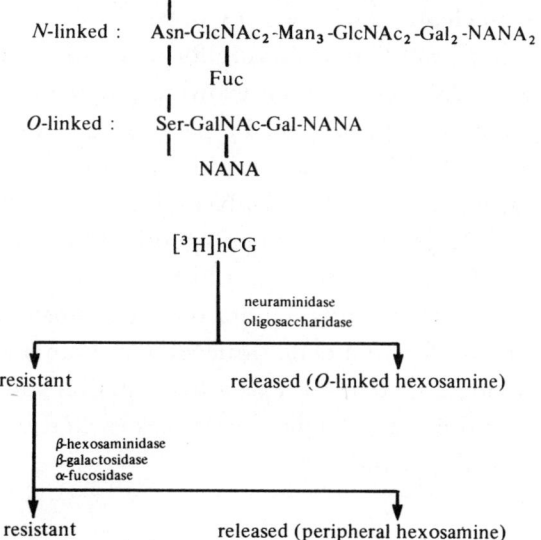

FIGURE 12. Procedure used to determine the distribution of the hexosamine units in the N- and O-linked chains of hCG.

of peripheral sugars, i.e. GlcNAc, Gal and NANA, occur in the Golgi complex. The fact that the timing of addition of GalNAc units to O-linked chains and of peripheral GlcNAc units to N-linked chains is identical suggests that, indeed, O-linked glycosylation also occurs in the Golgi complex, and is therefore not a co-translational process. At later time intervals the distribution pattern radically changes, as expected, because now newly synthesized hCG chains are appearing in the culture medium. Under these circumstances most of the label is found in GlcNAc units in the *N*-linked core. In fact, the distribution of the three types of labelled amino sugars is in agreement with the theoretical values for mature hCG. These results, as well as the experiments carried out in oviduct tissue slices *in vitro*, strongly support the idea that *O*-glycosylation is a post-translational process.

With reference to the scheme outlined in figure 1, it is of interest to consider why radically different mechanisms are used for the addition of *N*- and *O*-linked chains. Why does the attachment of *N*-linked chains involve the preassembly of the completed oligosaccharide on

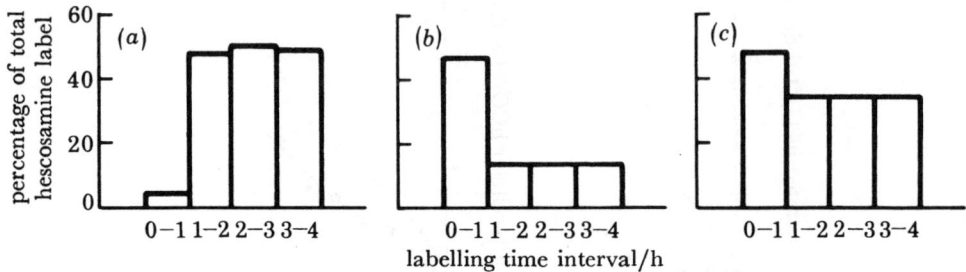

FIGURE 13. Distribution of GlcNAc and GalNAc units in secreted hCG:
(*a*) *N*-core GlcNAc; (*b*) *O*-GalNAc; (*c*) peripheral GlcNAc.

dolichyl pyrophosphate, whereas the synthesis of the *O*-linked chains involves the direct transfer of the sugars from sugar nucleotides without any preassembly process? One possible explanation is that both the assembly of an oligosaccharide for a *N*-linked chain containing 14 saccharide units and its transfer to an Asn residue on a growing polypeptide must be rapid processes that presumably are accomplished with fidelity. Prepackaging of the oligosaccharide on a dolichyl phosphate 'anchor' would facilitate such rapid transfer. The need for such a preassembly process is not obvious when one considers *O*-glycosylation, as well as the addition of peripheral sugars to *N*-linked chains. In these cases the polypeptides are complete and are being stored, at least temporarily, in the Golgi complex. In addition, these late steps involve the addition a relatively smaller number of sugar units. Under these circumstances the need for a preassembly process to ensure rapid transfer would be unnecessary. If this idea is correct, the function of dolichyl phosphate may merely be to serve as a hydrophobic anchor for the preassembly of the oligosaccharide chains, and not as a 'sugar carrier' as earlier postulated for lipid intermediates in the synthesis of bacterial glycans.

REGULATION OF GLYCOSYLATION

Finally, it is of interest to consider possible mechanisms for the regulation of protein glycosylation. Three possible factors that could serve a regulatory role in the glycosylation of proteins would be the availability of (*a*) sugar nucleotides, (*b*) dolichyl phosphate, and (*c*) nascent polypeptide chains with glycosylatable Asn sites. One indication that dolichyl phosphate might, at least in some systems, be a rate-limiting factor in glycosylation came from studies on the relation between glycoprotein synthesis and embryonic development in sea urchins (Carson & Lennarz 1979). In these studies we had observed that compactin, an inhibitor of HMG–CoA reductase, blocked dolichol synthesis and caused an arrest in development. However, when embryos were cultivated in the presence of compactin plus dolichol, glycosylation activities, as assessed by synthesis of Man-P-dolichol, oligosaccharide-PP-dolichol and N-linked glycoproteins, were not only restored but significantly exceeded control values. This finding, suggesting that under normal circumstances dolichyl phosphate may be rate-limiting, prompted us to examine other biological systems. As shown in table 2, experiments with the fibroblastic LM

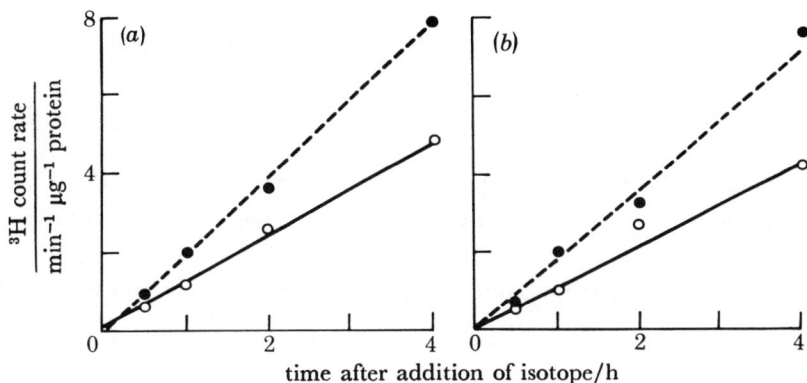

FIGURE 14. Stimulation of glycosylation of proteins in oviduct tissue slices by dolichyl phosphate: (a) [³H]Man; (b) [³H]GlcNAc; ●, with Dol-P; ○, without Dol-P.

TABLE 2. EFFECT OF EXOGENOUS DOLICHYL PHOSPHATE ON OLIGOSACCHARIDE-LIPID SYNTHESIS IN LM CELLS

	radioactive count in oligosaccharide-lipid	
	---	---
addition to culture medium	min⁻¹	
	10 min	20 min
none	6 500	14 000
Dol-P	17 000	38 000

TABLE 3. DOLICHYL PHOSPHATE SUPPLEMENTATION DOES NOT STIMULATE OVALBUMIN TRANSLATION OR GLYCOSYLATION

	radioactive count in anti-ovalbumin precipitate		
	---	---	---
	min⁻¹ mg⁻¹ protein		
addition to culture medium†...	[³H]mannose	[³⁵S]methionine	[³H]/[³⁵S]
control	70 305	13 730	5.1
plus dolichyl phosphate	68 243	12 406	5.5

† Oviduct tissue slices (50 mg wet mass) were preincubated with or without dolichyl phosphate (20 μg ml⁻¹).

cell line indicated that the synthesis of oligosaccharide-lipid *in vivo* was doubled by supplementation of the culture medium with dolichyl phosphate (Grant & Lennarz 1982). Similar findings were observed in a very different system, namely oviduct tissue slices (Carson *et al.* 1981). Synthesis of N-linked glycoproteins in such slices was measured by using radioactive Man or GlcNAc as precursors. As shown in figure 14, it was found that the presence of dolichyl phosphate in the culture medium consistently doubled or trebled the total N-linked glycoprotein synthesis. However, when we specifically measured the effect of added dolichyl phosphate on the synthesis of the major glycoprotein produced by oviduct, ovalbumin, it was found that neither its synthesis nor its glycosylation was stimulated by added dolichyl phosphate (table 3). Further study revealed that the stimulation observed was the result of increased glycosylation of a variety of as yet unidentified glycoproteins. For this reason, we turned to another tissue slice system, bovine pancreas, which synthesizes and secretes ribonuclease (Carson *et al.* 1981). Ribonuclease secreted from bovine pancreas under normal conditions is largely unglycosylated;

TABLE 4. DOLICHYL PHOSPHATE SUPPLEMENTATION STIMULATES RIBONUCLEASE

GLYCOSYLATION IN BOVINE PANCREAS TISSUE SLICES

	radioactive count in ribonuclease		
		min^{-1}	
addition to culture medium†...	[³H]mannose	[¹⁴C]amino acids	[³H]/[¹⁴C]
− dolichyl phosphate	78 097	5 168	15.1
+ dolichyl phosphate	409 050	6 650	61.5

† Bovine pancreas tissue slices (50 mg wet mass) in 5 ml medium were preincubated with or without dolichyl phosphate (20 μg ml⁻¹).

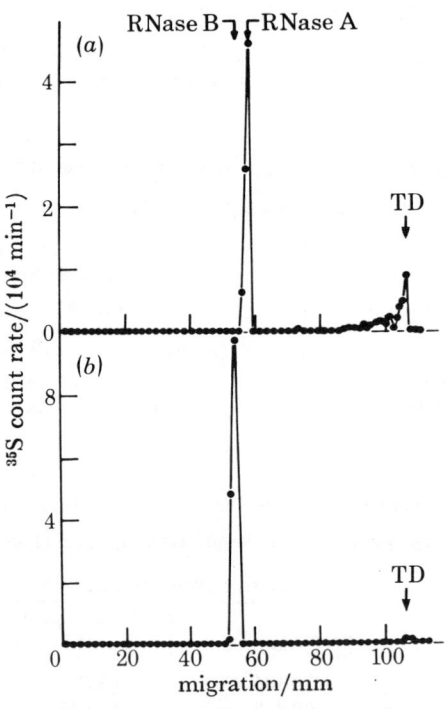

FIGURE 15. Sodium dodecyl sulphate polyacrylamide gel electrophoresis of ribonuclease secreted by pancreatic tissue slices incubated in the absence (*a*) or presence (*b*) of dolichyl phosphate.

only 10 % of the secreted chains are ribonuclease B, the glycosylated form of the enzyme. Using an affinity column to isolate ribonuclease secreted from the tissue slices, we assessed the affect of dolichyl phosphate in the culture medium on both mannose and amino acid incorporation. As shown in table 4, amino acid incorporation into secreted ribonuclease was unaffected by the presence of dolichyl phosphate, whereas incorporation of mannose was very markedly stimulated. To verify that enhanced glycosylation of the ribonuclease occurred in the presence of dolichyl phosphate, the product formed under these two conditions was examined by polyacrylamide gel electrophoresis. As shown in figure 15, it is clear that virtually all of the ribonuclease synthesized and secreted by the slices in the absence of dolichyl phosphate is ribonuclease A, the unglycosylated form. In contrast, in tissue sizes supplemented with dolichyl phosphate, most of the product is ribonuclease B, the glycosylated form. Final confirmation of this finding was accomplished by assessing the extent of glycosylation of ribonuclease as a function of the concentration of dolichyl phosphate in the culture medium. As shown in figure

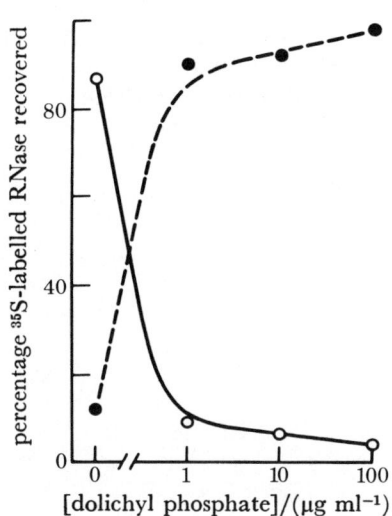

FIGURE 16. Effect of exogenous dolichyl phosphate on the synthesis of glycosylated ribonuclease in pancreatic tissue slices. ●, RNase bound to Con A–Sepharose; ○, RNase not bound.

16, in the absence of dolichyl phosphate most of the ribonuclease, as assessed by its ability to bind to Concanavalin A, is unglycosylated. In contrast, upon addition of even low levels of exogenous dolichyl phosphate most of the secreted ribonuclease is found to be glycosylated. Thus it is clear that, at least in the systems investigated so far, dolichyl phosphate can be a rate-limiting factor in glycoprotein synthesis. Whether or not sugar nucleotides and availability of glycosylatable polypeptide chains also serve as regulatory factors remains to be established.

The work reported from the author's laboratory was supported by grants from the National Institutes of Health (nos GM 21451 and GM 24821). It is a pleasure to acknowledge the efforts of my associates in these studies: D. D. Carson, B. J. Earles, J. Elting, C. G. Glabe, S. Grant, J. A. Hanover and G. R. Mintz.

REFERENCES

Ballas, L. M. & Arion, W. J. 1977 *J. biol. Chem.* **252**, 8512–8518.
Blobel, G. & Sabatini, D. C. 1970 *J. Cell Biol.* **45**, 130–145.
Carey, D. J. & Hirschberg, C. B. 1981 *J. biol. Chem.* **256**, 989–993.
Carson, D. D., Earles, B. J. & Lennarz, W. J. 1981 *J. biol. Chem.* **256**, 11552–11557.
Carson, D. D. & Lennarz, W. J. 1979 *Proc. natn. Acad. Sci. U.S.A.* **76**, 595–604.
Fleischer, B. 1981 *J. Cell Biol.* **89**, 246–255.
Glabe, C. G., Hanover, J. A. & Lennarz, W. J. 1980 *J. biol. Chem.* **255**, 9236–9242.
Grant, S. & Lennarz, W. J. 1982 (Submitted.)
Hanover, J. A., Elting, J., Mintz, G. R. & Lennarz, W. J. 1982 *J. biol. Chem.* (In the press.)
Hanover, J. A. & Lennarz, W. J. 1980 *J. biol. Chem.* **255**, 3600–3604.

Hanover, J. A. & Lennarz, W. J. 1982 *J. biol. Chem.* **257**, 2787–2794.

Hanover, J. A., Lennarz, W. J. & Young, J. D. 1980 *J. biol. Chem.* **255**, 6713–6716.

Jokinen, M., Gahmberg, C. G. & Andersson, L. C. 1979 *Nature, Lond.* **279**, 604–607.

Jokinen, M., Ulmanen, I., Andersson, L. C., Kaariainen, L. & Gahmberg, C. G. 1981 *Eur. J. Biochem.* **114**, 393–397.

Kim, Y. S., Perdomo, J. & Nordberg, J. 1971 *J. biol. Chem.* **246**, 5466–5476.

Ko, G. K. W. & Raghupathy, E. 1972 *Biochim. biophys. Acta* **264**, 129–143.

Nesbit, A. D., Saundry, R. H., Moir, A. J., Fothergill, L. A. & Fothergill, J. E. 1981 *Eur. J. Biochem.* **115**, 335–345.

Ruddon, R. W., Bryan, A. H., Hanson, C. A., Perini, F., Ceccorulli, L. M. & Peters, B. P. 1981 *J. biol. Chem.* **256**, 5189–5196.

Ruddon, R. W., Handon, C. A., Bryan, A. H., Rutterman, G. J., White, E. L., Perini, F., Meade, K. S. & Aldenderfer, P. H. 1980 *J. biol. Chem.* **255**, 1000–1007.

Strous, G. J. A. M. 1979 *Proc. natn. Acad. Sci. U.S.A.* **76**, 2694–2698.

White, D. A. & Speake, B. K. 1980 *Biochem. J.* **192**, 297–301.

Phil. Trans. R. Soc. Lond. B **300**, 145–159 (1982)
Printed in Great Britain

Oligosaccharide branching of glycoproteins: biosynthetic mechanisms and possible biological functions

By H. Schachter, S. Narasimhan, P. Gleeson, G. J. Vella
and I. Brockhausen

*Research Institute, Hospital for Sick Children, 555 University Avenue,
Toronto, Ontario, Canada M5G 1X8*

One of the most striking features of *N*- and *O*-glycosyl oligosaccharides and of lipid-linked oligosaccharides is the high degree of branching of these complex structures. Both proteins and nucleic acids are essentially linear structures and are synthesized by template mechanisms. The branched nature of complex carbohydrates dictates a totally different mechanism of biosynthetic control. Although there are undoubtedly many factors controlling this assembly (e.g. subcellular compartmentation, availability of substrates, cations), our laboratory has studied primarily the enzymatic factors that control the assembly of branched *N*-glycosyl (Asn-GlcNAc type) and *O*-glycosyl (Ser[Thr]-GalNAc type) oligosaccharides. There are three basic types of control points that appear to direct biosynthesis. (*a*) There may be two or more enzymes capable of acting on a single common substrate. Control at this juncture is exerted by the relative activities of these enzymes in a particular tissue. (*b*) Addition of a specific sugar to the growing oligosaccharide may shut off one or more subsequent enzyme steps, thereby 'freezing' the structure at a certain stage in its synthesis. (*c*) Progression of the pathway may be impossible until a certain key sugar residue is inserted into the growing oligosaccharide chain. Examples of all three types of control occur in the assembly of both *N*- and *O*-glycosyl oligosaccharides. This paper discusses our work on the *N*-acetylglucosaminyltransferases, which initiate branches in *N*-glycosyl oligosaccharides, as well as some studies on glycosyltransferases that control the assembly of the four basic Ser(Thr)-GalNAc cores. Important features at all stages of control are the three-dimensional shape of the oligosaccharide, the effect of certain key sugar residues on this three-dimensional shape and the stereochemistry of the interaction of oligosaccharides with proteins. From a functional point of view, protein–oligosaccharide interaction is of vital importance not only to enzyme control mechanisms but to a variety of biological problems such as malignancy and cell–cell interactions, differentiation and development, and susceptibility of cells to hormones, drugs and toxins.

Introduction

One of the most consistent biochemical correlations with malignant transformation is a relative increase in the release of high molecular mass sialic acid-rich glycopeptide material from the cell surface by proteolytic cleavage (Warren *et al.* 1978). Ogata *et al.* (1976) first suggested that this material contained highly branched (tri- and tetra-antennary) Asn-linked oligosaccharides. Proof for this hypothesis was obtained by Takasaki *et al.* (1980), who found that transformation of BHK cells with polyoma virus led to a relative decrease in bi-antennary oligosaccharides, a relative increase in tetra-antennary oligosaccharides and the appearance of novel penta- and hexa-antennary structures. However, the correlation between degree of branching and malignant transformation has exceptions (Warren *et al.* 1978; Rachesky *et al.* 1982; Cossu *et al.* 1982). Our laboratory has for several years studied the enzymic mechanisms responsible for the branching of both *N*-glycosyl and *O*-glycosyl oligosaccharides. This topic is of interest not only because of the possible association of branched structures with malignancy but also because the degree of branching varies in other situations and the functional significance of these variations remains obscure.

STRUCTURAL CONSIDERATIONS
Asn-GlcNAc oligosaccharides

Asn-GlcNAc oligosaccharides appear to have a common core:

$$\begin{array}{c} Man\alpha1\text{-}3 \\ \\ Man\alpha1\text{-}6 \end{array} \Big\rangle Man\beta1\text{-}4GlcNAc\beta1\text{-}4GlcNAc\text{-}Asn\text{-}X.$$

Relatively primitive organisms (e.g. yeasts) have only the 'high-mannose' type of structure, which contains a large number of Man residues attached to this common core. More advanced organisms (e.g. birds and mammals) have retained this high-mannose structure (although with far fewer Man residues than are found in yeasts) and have developed in addition a more 'complex' series of structures. There are now at least ten classes of *N*-glycosyl oligosaccharides (Carver & Grey 1981; Turco *et al.* 1980) and more may eventually be discovered. These classes are: bi-antennary complex (I), tri-antennary complex (II), tetra-antennary complex (III), bisected bi-antennary complex (IV), bisected tri-antennary complex (V), high mannose (VI), bi-antennary hybrid (VII), bisected bi-antennary hybrid (VIII), bisected tri-antennary hybrid (IX) and polylactosaminoglycan or erythroglycan (X). Figure 1 shows examples of some of these structures.

Ser(Thr)-GalNAc oligosaccharides

Ser(Thr)-GalNAc oligosaccharides are even more diverse than the *N*-glycosyl oligosaccharides. The simplest structures (found, for example, in ovine submaxillary mucin) are GalNAc-Ser(Thr)-X and sialylα2-6GalNAc-Ser(Thr)-X. More complex structures have recently been classified into four core types (Schachter & Williams 1982):

type 1 core, Galβ1-3GalNAc-Ser(Thr)-X;

type 2 core, Galβ1-3(GlcNAcβ1-6)GalNAc-Ser(Thr)-X;

type 3 core, GlcNAcβ1-3GalNAc-Ser(Thr)-X;

type 4 core, GlcNAcβ1-3(GlcNAcβ1-6)GalNAc-Ser(Thr)-X.

These cores can then be elongated into a variety of different branched structures, some of which carry human blood group antigenic determinants such as A, B, H, Lewis[a], Lewis[b] and Ii.

BIOSYNTHESIS OF Asn-GlcNAc HIGH-MANNOSE STRUCTURES

The preceding papers have dealt in detail with the early steps in the assembly of *N*-glycosyl oligosaccharides. Biosynthesis begins within the rough endoplasmic reticulum with transfer of $(Glc)_3(Man)_9(GlcNAc)_2$ from dolichyl pyrophosphate-oligosaccharide to nascent polypeptide. Processing within the rough endoplasmic reticulum removes the 3Glc residues; this is followed

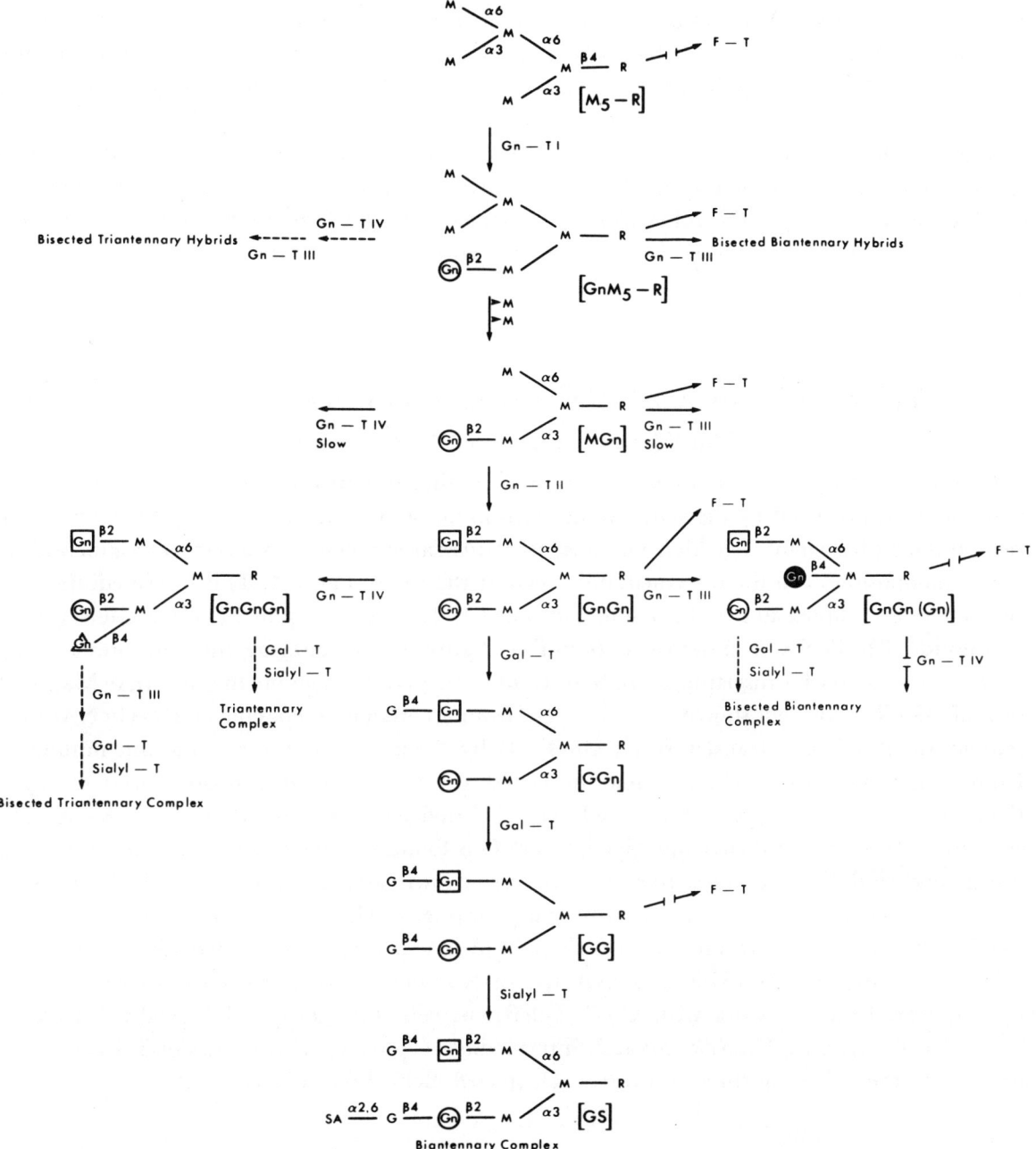

FIGURE 1. Golgi-localized pathway for the assembly of *N*-glycosyl hybrid and complex oligosaccharides. The pathway is discussed in detail in the text. Hybrid oligosaccharides carry Man residues on the Manα1-6- arm of the core and antennae (usually GlcNAc- or Galβ1-4GlcNAc-) on the Manα1-3- arm. Hybrid oligosaccharides may be non-bisected (like structure GnM$_5$-R) but are usually bisected by a GlcNAc residue linked β1-4- to the β-linked Man residue. Abbreviations: M, Man; Gn, GlcNAc; F, Fuc; G, Gal; SA, sialyl; R, GlcNAcβ1-4GlcNAc-Asn-X; T, transferase. Glycopeptide nomenclature: complex oligosaccharides containing only 3Man residues are named according to the sugars at the non-reducing termini of the antennae; the sugar on the Manα1-6- arm is named first. Order of sugar addition: the order of GlcNAc addition is discussed in the text. Immunoglobulins isolated from human multiple myeloma serum have been shown to carry the structures GGn and GS (rather than GnG and SG) suggesting that Gal-T prefers the Manα1-6- arm and sialyl-T prefers the Manα1-3- arm. Enzymic evidence in support of this ordered addition has been obtained (Van den Eijnden *et al.* 1980; Rao & Mendicino 1978).

[39]

by the removal, partly in the rough endoplasmic reticulum but mainly within the Golgi apparatus, of four α-2-linked Man residues, to yield a key intermediate (shown at the top of figure 1) containing 5Man residues and 2GlcNAc residues. This structure and a similar compound containing only 4Man residues are the smallest high-mannose structures found in avian and mammalian glycoproteins. The largest high-mannose structure found in higher organisms contains 9Man residues, although yeast glycoproteins may contain much larger oligosaccharides. For some protein-bound oligosaccharides, the synthetic path goes no further than the high-mannose stage. For others, unknown factors (possibly the amino acid sequence near the relevant Asn residue) direct the synthetic path towards complex oligosaccharides (see below).

BIOSYNTHESIS OF Asn-GlcNAc COMPLEX AND HYBRID STRUCTURES

UDP-GlcNAc: α-D-mannoside β2-GlcNAc-transferase I

Complex and hybrid structures occur only in higher organisms, and the synthesis of these compounds is preceded by the synthesis of high-mannose structures, i.e. the synthetic pathway recapitulates phylogeny. The high-mannose structures are processed by a series of α-glucosidases and α-mannosidases to the intermediate shown at the top of figure 1. This intermediate is the precursor of complex and hybrid oligosaccharides. It is acted upon by UDP-GlcNAc: α-D-mannoside β2-GlcNAc-transferase I (Gn-T I, figure 1), an enzyme that initiates the first branch or antenna by inserting a GlcNAc residue in β1-2 linkage on the terminal Manα1-3-residue. Gn-T I was first shown to be different from an analogous enzyme (UDP-GlcNAc: α-D-mannoside β2-GlcNAc-transferase II, Gn-T II) by the isolation of lectin-resistant mutants of Chinese hamster ovary cells deficient in Gn-T I (Stanley 1980; Narasimhan et al. 1977; Gottlieb et al. 1974, 1975). When such Gn-T I-deficient cells are infected with vesicular stomatitis virus, the viral envelope glycoprotein G is found to have oligosaccharides containing 5Man and 2GlcNAc residues, like the structure shown at the top of figure 1 (Tabas et al. 1978). Normal glycoprotein G has only complex oligosaccharide structures, indicating that Gn-T I-deficient cells are unable to make complex structures. Uninfected Gn-T I-deficient cells also accumulate the (Man)$_5$(GlcNAc)$_2$-Asn-X structure (Li & Kornfeld 1978).

It appears from the work with Gn-T I-deficient cells that the physiological substrate for Gn-T I is the (Man)$_5$(GlcNAc)$_2$-Asn-X intermediate (figure 1). However, Gn-T I can act on any one of the following three substrates (Harpaz & Schachter 1980a):

Manα1-6
↘
and   Manβ1-4GlcNAcβ1-4GlcNAc-Asn-X.
↗
Manα1-3

The action of Gn-T I on $(Man)_3(GlcNAc)_2$-Asn-X is the major physiological path in an interesting mouse lymphoma mutant that is unable to make dolichyl monophosphate mannose (Kornfeld *et al.* 1979; Chapman *et al.* 1980), and it may be the major synthetic path in normal cells under certain conditions (Rearick *et al.* 1981).

Johnston *et al.* (1966, 1973) first reported the presence in goat colostrum of a GlcNAc-transferase acting on α_1-acid glycoprotein pretreated with neuraminidase, β-galactosidase and β-*N*-acetylglucosaminidase. A similar enzyme has been found in many tissues (Schachter & Roseman 1980) of the rat, pig, guinea-pig and human and has been found to be localized in the Golgi apparatus in rat liver. The high molecular mass acceptor detects both Gn-T I and II, and specific glycopeptide acceptors must be used to assay Gn-T I and II independently (Harpaz & Schachter 1980*a*). Gn-T I has been purified from bovine colostrum (Harpaz & Schachter 1980*a*), rabbit liver (Oppenheimer & Hill 1981) and pig liver and tracheal mucosa (Oppenheimer *et al.* 1981).

As is evident from figure 1, Gn-T I controls the entry of the synthetic pathway towards complex and hybrid structures. Several enzymes can act on the product of Gn-T I but not on the substrate, i.e. Gn-T I-dependent α-3/6-mannosidase(s), GDP-Fuc: β-*N*-acetylglucosaminide (Fuc to Asn-linked GlcNAc) α6-Fuc-transferase (α6-Fuc-T) and probably UDP-GlcNAc: glycopeptide β4-GlcNAc-transferase III (Gn-T III). Lack of action of these three enzymes explains, respectively, why there is no complex oligosaccharide synthesis, no fucose incorporation and no hybrid oligosaccharide synthesis until Gn-T I has acted. This is an example of a control point in which the addition of a single critical sugar residue allows previously impossible reactions to occur.

The product of Gn-T I activity is

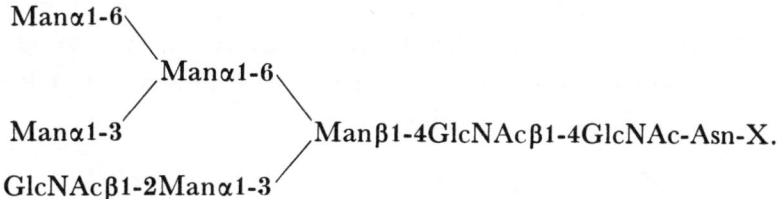

Gn-T I appears to be specific for the terminal Manα1-3- residue of the core because it does not attack the other more peripheral terminal Manα1-3- residue of the substrate. Further, Gn-T I acts on the Manα1-3- terminus of $(Man)_3(GlcNAc)_2$-Asn-X and not on the equally available Manα1-6- terminus. The smallest effective substrate for Gn-T I is the trisaccharide Manα1-3Manβ1-4GlcNAc (Harpaz & Schachter 1980*a*).

Gn-T I-dependent α-3/6-mannosidase(s)

The product of Gn-T I sits at a cross roads, i.e. it can be acted on by at least three (and probably four) different enzymes (figure 1). Although we have shown (Longmore & Schachter 1982) that α6-Fuc-T can act at this point to form a fucosylated non-bisected bi-antennary hybrid oligosaccharide, the natural occurrence of such a structure has not yet been demonstrated. Another possible pathway is directed towards the formation of hybrid structures via Gn-T III and Gn-T IV (this is discussed below). However, the most common fate for the product of Gn-T I (the 'main line' path) is the formation of complex oligosaccharides, containing 3Man residues, via the action of an unusual glycosidase that we have named Gn-T I-dependent α-3/6-mannosidase(s) (figure 1). This Golgi-localized enzyme or enzymes will act only on the product (but not on the substrate) of Gn-T I (Tabas & Kornfeld 1978; Harpaz & Schachter 1980b). The enzyme has been purified from rat liver Golgi-rich membranes (Tulsiani et al. 1981, 1982). Hen oviduct is rich in Gn-T III (Narasimhan et al. 1981a, b) but has relatively little Gn-T I-dependent α-3/6-mannosidase(s) activity. It is therefore not surprising to find that ovalbumin contains a large amount of bisected hybrid oligosaccharides (see below). Rat and pig liver Golgi-rich membranes have more Gn-T I-dependent α-3/6-mannosidase(s) activity than Gn-T III (unpublished data) activity and presumably make primarily complex oligosaccharides.

UDP-GlcNAc : α-D-mannoside β2-GlcNAc-transferase II

Lectin-resistant cells completely deficient in Gn-T I contain normal levels of UDP-GlcNAc: α-D-mannoside β2-GlcNAc-transferase II (Gn-T II), an enzyme that initiates the second branch or antenna on the Manα1-6- arm of the core (figure 1). Although Gn-T II has not yet been purified, it has been obtained free of Gn-T I from bovine colostrum (Harpaz & Schachter 1980a), pig liver and tracheal mucosa (Oppenheimer et al. 1981) and hen oviduct (unpublished data). The only effective substrate for Gn-T II is the product of Gn-T I-dependent α-3/6-mannosidase(s), i.e. glycopeptide MGn (figure 1). At least four enzymes can act on glycopeptide MGn, i.e. Gn-T II, Gn-T III, Gn-T IV and α6-Fuc-T (figure 1). Gn-T III and Gn-T IV, however, act relatively slowly on this substrate (table 1), and the formation of GnGn (with or without Fuc) via Gn-T II is no doubt the major fate of MGn. Gn-T II cannot act on Manα1-3(Manα1-6)Manβ1-4GlcNAcβ1-4GlcNAc-Asn-X, indicating that the prior action of Gn-T I is essential for its activity. Further, the presence of a bisecting GlcNAc prevents Gn-T II action (Harpaz & Schachter 1980a).

UDP-GlcNAc : glycopeptide β4-GlcNAc-transferase III

(a) Hybrid oligosaccharide formation

Some oligosaccharides are bisected by a GlcNAc residue linked β1-4 to the β-linked Man residue of the core. Tissues like hen oviduct, with a relative preponderance of Gn-T III over Gn-T I-dependent α-3/6-mannosidase(s), will insert a bisecting GlcNAc residue immediately after Gn-T I action (figure 1). This has a most interesting effect because the presence of a bisecting GlcNAc shuts off at least four different enzymes, i.e. Gn-T II, Gn-T IV, Gn-T I-dependent α-3/6-mannosidase(s) and α6-Fuc-T. The effect of Gn-T III action immediately after Gn-T I action is to 'freeze' oligosaccharide synthesis at the bisected bi-antennary hybrid stage. Synthesis stops except for elongation of the single antenna (initiated by Gn-T I) by addition of a Gal (and possibly a sialyl) residue. Sialylated hybrid structures have recently been found in lysosomal hydrolases.

Hen oviduct membranes are also rich in Gn-T IV (see below) and Gn-T IV may act on the product of Gn-T I before Gn-T III acts, thereby resulting in bisected tri-antennary hybrids.

Bovine rhodopsin (Liang *et al.* 1979) has been shown to contain glycopeptide MGn (figure 1) as well as the two non-bisected bi-antennary hybrid structures formed by the actions of Gn-T I and Gn-T I-dependent α-3/6-mannosidase(s) respectively (figure 1). The reason for the presence of these structures in rhodopsin is not known but interference with the actions of the mannosidase and Gn-T II is possible.

TABLE 1. GLYCOSYLTRANSFERASE SUBSTRATE SPECIFICITIES

(Hen oviduct membranes were used for studies with GlcNAc-transferase III (Gn-T III) at pH 5.7 and 7.0 (Narasimhan *et al.* 1981 *a, b*), and GlcNAc-transferase IV (Gn-T IV) at pH 7.0 (Gleeson *et al.* 1982 *a, b*). Golgi-enriched membranes from pig liver were used for the studies on GDP-Fuc: β-*N*-acetylglucosaminide (Fuc to Asn-linked GlcNAc) α1-6-fucosyltransferase (Fuc-T) (Longmore & Schachter 1982). These membranes are enriched 30-fold in Fuc-T relative to homogenate. One enzyme unit is 1 μmol min^{-1}.)

	specific activity (μunits mg^{-1})			
substrate†	Gn-T III, pH 5.7	Gn-T III, pH 7.0	Gn-T IV, pH 7.0	Fuc-T, pH 8.0
GnGn (+Fuc)	85	73	83	—‡
GnGn (−Fuc)	—	75	95	1800
MGn (+Fuc)	⩽13	10	18	0
MGn (−Fuc)	—	—	—	3100
MM (+Fuc)	⩽7	⩽16	⩽16	0
MM (−Fuc)	—	—	—	0
GGn (+Fuc)	—	48	30	—
GnG (+Fuc)	—	⩽5	⩽5	—
GG (−Fuc)	—	⩽8	⩽8	0§
SS (−Fuc)	—	0	0	0
GnGn(Gn) (−Fuc)	—	—	⩽3	0
MGn(Gn) (−Fuc)	—	—	—	0

† Structures of glycopeptides GnGn, MGn, GGn, GG and GnGn(Gn) are shown in figure 1. Glycopeptide MM is Manα1-6(Manα1-3)Manβ1-4GlcNAcβ1-4(Fucα1-6)GlcNAc-Asn-X. GnG is similar to GGn except that the Gal residue is on the Manα1-3- arm instead of the Manα1-6- arm. SS is fully sialylated bi-antennary complex glycopeptide with both sialyl residues linked α2-6-. The glycopeptides are named according to the terminal sugar residues: M, Man; Gn, GlcNAc; G, Gal; S, sialyl; (Gn), bisecting GlcNAc.

‡ Not done, or not relevant.

§ Based on work with rat liver Golgi membranes that lack fucosyltransferase activities towards Galβ1-4GlcNAc-terminated acceptors (Munro *et al.* 1975).

(b) Bisected complex oligosaccharide formation

Bisected complex oligosaccharides are relatively uncommon but have been found in several human multiple myeloma immunoglobulins (bisected bi-antennary complex oligosaccharides) and in ovotransferrin (bisected tri-antennary complex structures). Such structures cannot be formed by the action of Gn-T III at the 5-Man stage (figure 1), suggesting that there must be another point of entry for Gn-T III (Harpaz & Schachter 1980 *b*). Gn-T III activity acting on glycopeptide GnGn (figure 1) has indeed been found in hen oviduct membranes (Narasimhan *et al.* 1981 *a, b*). It is not known whether the Gn-T III activities acting at the 3-Man and 5-Man stages are one and the same enzyme.

The substrate specificity of Gn-T III is shown in table 1. It is evident that there is a relatively narrow 'window' of action (at glycopeptides MGn, GnGn and GGn; figure 1). As soon as both arms are galactosylated, Gn-T III action is no longer possible. As mentioned above, the introduction of a bisecting GlcNAc residue prevents further action of Gn-T IV. Thus if Gn-T III acts first, the synthetic pathway is 'frozen' into synthesis of bisected bi-antennary complex oligosaccharides (figure 1). If Gn-T IV acts before Gn-T III, synthesis of tri- and tetra-antennary oligosaccharides is possible. Glycopeptide GnGn is therefore at yet another important cross roads, leading into at least three different pathways, i.e. non-bisected bi-antennary complex, bisected bi-antennary complex, and tri- and/or tetra-antennary complex oligosaccharides; the last may remain unbisected or may become bisected via the action of Gn-T III.

UDP-GlcNAc : glycopeptide GnGn β4-GlcNAc-transferase IV

Hen oviduct membranes are rich in UDP-GlcNAc:glycopeptide GnGn β4-GlcNAc-transferase IV (Gn-T IV), which initiates the third branch or antenna by the addition of a GlcNAc residue in β1-4- linkage to the Manα1-3- arm of glycopeptide GnGn (figure 1) (Gleeson *et al.* 1982 *a, b*). The substrate specificity of Gn-T IV is shown in table 1. It shows the same narrow 'window' of specificity as Gn-T III. Thus highly branched (tri- and tetra-antennary) oligosaccharides, such as are believed to be present at increased levels in transformed cells, result presumably from a relative increase of Gn-T IV over both UDP-Gal:GlcNAc β4-galactosyltransferase (leading to bi-antennary complex structures) and Gn-T III (leading to bisected bi-antennary complex oligosaccharides). Prior action of Gn-T III prevents Gn-T IV action and directs the pathway into bisected bi-antennary complex structures. Galactosylation of both arms also prevents Gn-T IV action leading to bi-antennary complex oligosaccharides.

GDP-Fuc: β-N-acetylglucosaminide (Fuc to Asn-linked GlcNAc) α6-fucosyltransferase

The enzyme that attaches Fuc in α1-6- linkage to the Asn-linked GlcNAc of the core can act at various points in the synthetic scheme (Longmore & Schachter 1982) (figure 1; table 1). The enzyme requires the presence of the GlcNAc incorporated by Gn-T I, but is prevented from acting by the incorporation of a bisecting GlcNAc by Gn-T III. Galactosylation of the antennae prevents α6-Fuc-T action.

BIOSYNTHESIS OF Ser(Thr)-GalNAc OLIGOSACCHARIDES

No attempt will be made here to cover this complex area (see Schachter & Williams (1982) for a recent review). We shall briefly mention three examples of control points in the O-glycosyl oligosaccharide synthetic pathways.

Ovine and porcine submaxillary mucins

Ovine submaxillary mucin (o.s.m.) contains mainly GalNAc-Ser(Thr)-X and sialylα2-6GalNAc-Ser(Thr)-X oligosaccharides with small amounts (less than 2%) of Galβ1-3GalNAc-Ser(Thr)-X and fucosylated oligosaccharides. Porcine submaxillary mucin (p.s.m.) contains qualitatively similar structures but a much larger proportion of tetrasaccharides (and, in some pigs, pentasaccharides). The key control point (figure 2) occurs at GalNAc-Ser(Thr)-X, an intermediate that can be acted upon by two enzymes, UDP-Gal:GalNAc-mucin β1-3-Gal-transferase (mucin β3-Gal-T) and CMP-sialic acid:GalNAc-mucin α2-6-sialyltransferase (mucin α6-sialyl-T). Ovine submaxillary glands have a higher α6-sialyl-T/β3-Gal-T ratio than porcine glands, which leads to the sialyl-α2-6GalNAc- structure. The presence of the sialyl residue prevents β3-Gal-T action and 'freezes' the structure in the o.s.m. format. Conversely, in pig glands, β3-Gal-T functions more rapidly than α6-sialyl-T, which leads to the larger structures characteristic of p.s.m.

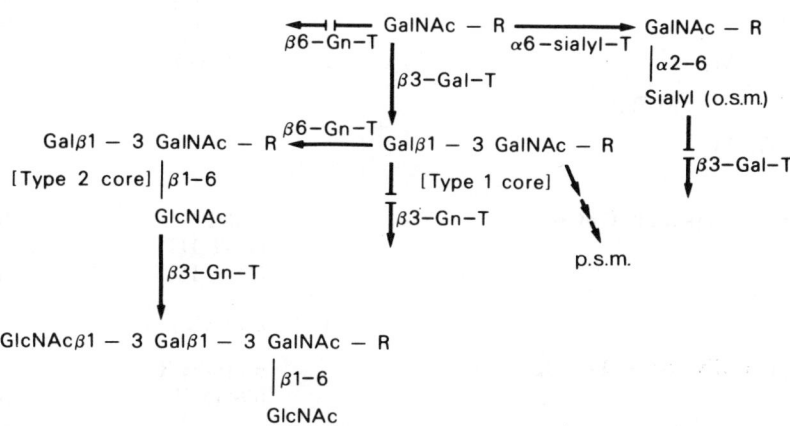

FIGURE 2. Paths involved in the synthesis of O-glycosyl oligosaccharides with type 1 and type 2 cores. Abbreviations: see text for the full names of the various transferases; o.s.m., ovine submaxillary mucin; p.s.m., porcine submaxillary mucin; R, Ser(Thr)-X.

Type 2 core synthesis and elongation

The synthesis of type 2 core requires the action of UDP-GlcNAc:Galβ1-3GalNAc-mucin (GlcNAc to GalNAc) β1-6-GlcNAc-transferase (mucin β6-Gn-T) (Williams & Schachter 1980; Williams et al. 1980). Mucin β6-Gn-T does not act on GalNAc-mucin but requires the prior insertion of the Galβ1-3- residue by mucin β3-Gal-T(figure 2). Galβ1-3GalNAc-mucin is in fact a key intermediate in O-glycosyl oligosaccharide synthesis (Schachter & Williams 1982). The following enzymes compete for this intermediate in various tissues: (a) elongation of type 1 core by the addition of sialic acid (α2-3- to Gal or α2-6- to GalNAc), or of fucose (α1-2- to Gal), or of GlcNAc in various linkages to Gal, and (b) synthesis of type 2 core via mucin β6-Gn-T.

We have recently shown (unpublished data) the elongation in vitro of type 2 core by a novel enzyme present in porcine gastric mucosa and in other mucus-secreting tissues, i.e. UDP-GlcNAc:Galβ1-3(GlcNAcβ1-6)GalNAc-R β1-3-GlcNAc-transferase (mucin β3-Gn-T). This enzyme does not appear to act on Galβ1-3-GalNAc-R but requires the prior insertion of GlcNAcβ1-6- by mucin β6-Gn-T. The type 2 core structure Galβ1-3(GlcNAcβ1-6)GalNAc-R is also at a cross roads in O-glycosyl oligosaccharide synthesis (Schachter & Williams 1982) and can be elongated not only by the addition of GlcNAc in β1-3- linkage to Gal by β3-Gn-T but also by the addition of Gal or Fuc to the GlcNAc residue of the core.

TABLE 2. CONTROL POINTS DURING OLIGOSACCHARIDE BIOSYNTHESIS

type	substrate (figure 1 or 2)	key residue	enzymes active or inactive towards the substrate	synthetic fate†
			active enzymes:	
(a) competing enzymes	(i) product of Gn-T I: 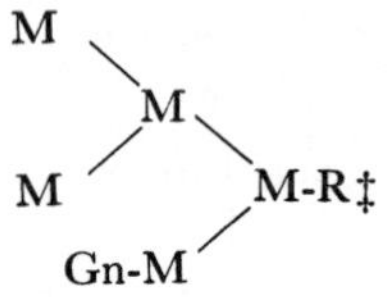	—	Gn-T I-dependent α3/6-mannosidase	all com
			Gn-T III	bis/bi/hy
			Gn-T IV	bis/tri/hy
			α6-Fuc-T	—
	(ii) glycopeptide GnGn	—	Gal-T	bi/com
			Gn-T III	bis/bi/com
			Gn-T IV	tri/tetra/com (+/− bis)
			α6-Fuc-T	—
	(iii) GalNAc-Ser(Thr)-X	—	α6-sialyl-T	o.s.m.§
			β3-Gal-T	p.s.m.¶
			inactive enzymes:	
(b) key residue 'freezes' pathway	(i) 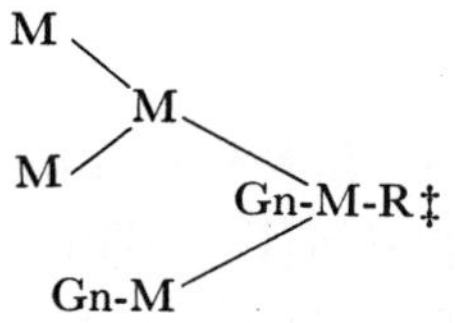	bisecting GlcNAc	Gn-T I-dependent α3/6-mannosidase	bis/bi/hy
			Gn-T IV	bis/bi/hy
			α6-Fuc-T	bis/bi/hy
	(ii) glycopeptide GnGn(Gn)	bisecting GlcNAc	Gn-T IV	bis/bi/com
			α6-Fuc-T	bis/bi/com
	(iii) sialylα2-6-GalNAc-Ser(Thr)-X	sialyl residue	β3-Gal-T	o.s.m.
			active enzymes:	
(c) key residue 'channels' pathway	(i) product of Gn-T I (see above)	GlcNAc inserted by Gn-T I	same as (a) (i) above	same as (a) (i) above
	(ii) Galβ1-3-GalNAc-Ser(Thr)-X	Galβ1-3-	β6-GlcNAc-T	type 2 core synthesis
	(iii) Galβ1-3-(GlcNAcβ1-6-)GalNAc-Ser(Thr)-X	GlcNAcβ1-6-	β3-GlcNAc-T	elongation of type 2 core

Structures shown in substrate column:

(a)(i):
```
M
 \
  M
 /    \
M      M-R‡
 \    /
  Gn-M
```

(b)(i):
```
M
 \
  M
 /      \
M        Gn-M-R‡
 \      /
  Gn-M
```

† Oligosaccharide products of the pathway. Abbreviations: bis, bisecting; bi, bi-antennary; tri, tri-antennary; tetra, tetra-antennary; com, complex; hy, hybrid.

‡ See figure 1 for full structures. M, Man; Gn, GlcNAc; R, GlcNAcβ1-4GlcNAc-Asn-X.

§ O.s.m., ovine submaxillary mucin.

¶ P.s.m., porcine submaxillary mucin.

[46]

Proteins and nucleic acids are essentially linear structures and are synthesized by a template mechanism. Oligosaccharides, being highly branched, cannot be assembled in this manner. Many factors undoubtedly play a role in oligosaccharide assembly, e.g. glycosyltransferase substrate specificity, the organization of the endomembrane assembly systems, the availability

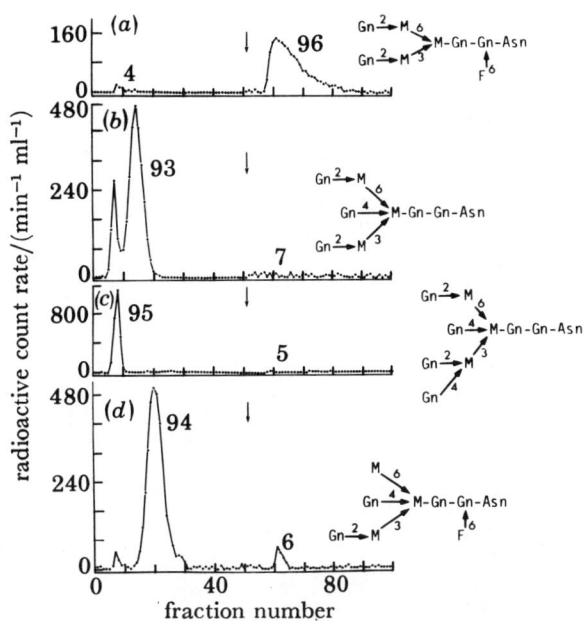

FIGURE 3. Radioactive glycopeptide standards were analysed on columns of Concanavalin A–Sepharose (0.7 cm × 15 cm) at pH 7.0. Arrows indicate start of elution with buffered 0.1 M methyl α-D-glucopyranoside. The figures near the peaks indicate the percentage radioactivity recovered in the peak relative to total radioactivity recovered from the column. Fraction size was 1.0 ml. Abbreviations in the structural formulae: M, Man; Gn, GlcNAc; F, Fuc. (a) N-[14C]acetyl-glycopeptide GnGn. A radioactive count of 1940 min⁻¹ was loaded; recovery was 100%. GnGn adheres firmly to the column and elutes in a typically broad peak with methyl α-D-glucopyranoside. (b) N-[14C]acetyl-glycopeptide GnGn(Gn). Radioactive load, 3550 min⁻¹; recovery, 95%. The introduction of a bisecting GlcNAc into GnGn to form GnGn(Gn) results in weakened binding to the lectin column. GnGn(Gn) comes through the column in retarded fashion (fractions 10–20). The material passing through totally unretarded (fractions 5–10) is a contaminant in the preparation (see Longmore & Schachter 1982). (c) N-[14C]acetyl-glycopeptide GnGnGn(Gn). Radioactive load, 3060 min⁻¹; recovery, 98%. Addition of a third antenna to GnGn(Gn) to form GnGnGn(Gn) results in a total lack of binding to the lectin column. This property allows the separation of the product of GlcNAc-transferase III (retarded) from the product of transferase IV (unretarded). (d) N-[14C]acetyl-glycopeptide MGn(Gn); radioactive load, 3860 min⁻¹; recovery, 95%. Glycopeptide MGn (see figure 1 for structure) binds tightly to Concanavalin A–Sepharose (data not shown). The introduction of a bisecting GlcNAc to MGn to form MGn(Gn) results in weakened binding, as indicated in the figure.

of substrates (acceptors and activated sugar donors), and the presence of cations and other cofactors. This discussion will deal with glycosyltransferase substrate specificity. A careful analysis of the synthetic paths discussed above indicates three types of control points: (a) two or more enzymes can compete for a common intermediate, (b) the addition of a 'key residue' to the growing oligosaccharide turns off further synthesis and 'freezes' the structure at a certain assembly stage, and (c) progression of the pathway is impossible unless a certain 'key residue' is inserted into the growing oligosaccharide. Table 2 summarizes examples of all three types of control points.

Control by competition seems to be exerted simply by the relative activities of various enzymes in a particular tissue. Thus hen oviduct is rich in Gn-T III and Gn-T IV relative to Gn-T I-dependent α-3/6-mannosidase(s), whereas in rat and pig liver exactly the reverse holds; the consequences have been discussed above. Similarly, ovine submaxillary glands have a higher α6-sialyl-T/β3-Gal-T ratio than porcine glands, leading to the structures typical of

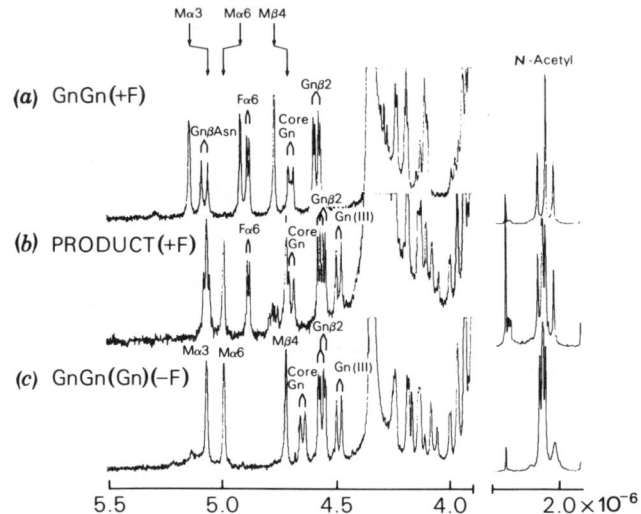

FIGURE 4. Anomeric and N-acetyl proton regions of the high resolution proton nuclear magnetic resonance (n.m.r.) spectra of glycopeptides. (a) Glycopeptide GnGn (with Fuc) prepared from human immunoglobulin G. (b) Glycopeptide GnGn(Gn), the product of GlcNAc-transferase III action on this GnGn preparation. (c) GnGn(Gn), without Fuc, isolated from hen ovalbumin. Glycopeptide nomenclature and structures are indicated in figure 1. The identification of the various signals is shown in the figure. Gn(III) indicates the signal for a bisecting GlcNAc residue. Note the dramatic shifts in the signals of all three Man residues caused by the presence of a bisecting GlcNAc residue. The signals for all four GlcNAc residues of GnGn and for Fuc are not appreciably altered by the introduction of a bisecting GlcNAc. The addition of a Fuc residue results in the appearance of its anomeric hydrogen signal (shown as Fα6) and in a downfield shift of the core GlcNAc (the residue attached to GlcNAc, not to Asn). It is curious that the signal for the Asn-linked GlcNAc (indicated by GnβAsn), hidden under the Manα1-3- signal, is not affected by attachment of a Fuc residue to the Asn-linked GlcNAc. The N-acetyl hydrogen resonances have been scaled down relative to the anomeric hydrogen signals. Addition of a bisecting GlcNAc residue results in an additional signal in the N-acetyl region.

o.s.m. and p.s.m. respectively. The other two types of control points depend on 'key residues' that either turn off or permit enzyme action. One key residue deserves particular attention because it acts at many points: the bisecting GlcNAc inserted by Gn-T III.

The presence of a bisecting GlcNAc turns off at least four enzymes (Gn-T II and IV, Gn-T I-dependent α-3/6-mannosidase(s), and α6-Fuc-T), suggesting a distortion in the three-dimensional structure of the oligosaccharide that prevents proper interaction between it and the various enzyme proteins. We have other evidence of interference in oligosaccharide–protein interaction by a bisecting GlcNAc residue. We have observed that the insertion of a bisecting GlcNAc into an oligosaccharide weakens its interaction with Concanavalin A–Sepharose (figure 3). This property has in fact proved useful in assaying Gn-T III since the substrate (glycopeptide GnGn, figure 3a) adheres firmly to the lectin column and requires 0.1 M methyl α-D-glucopyranoside for elution, whereas the product GnGn(Gn) adheres very weakly to the lectin (figure 3*b*).

The introduction of a bisecting GlcNAc also causes dramatic changes in the high-resolution proton nuclear magnetic resonance (n.m.r.) spectra of an oligosaccharide (Carver & Grey 1981). Figure 4(*a, b, c*) shows the anomeric hydrogen regions of the spectra of glycopeptides GnGn (with Fuc), GnGn(Gn) (with Fuc) and GnGn(Gn) (without Fuc) (see figure 1 for structures). The effect of a bisecting GlcNAc is seen by comparing figure 4*a* with figure 4*b*. Not only is there a dramatic shift in the signal for the β-linked Man (to which the bisecting GlcNAc is attached), but there are equally dramatic shifts in the signals for the other two Man residues. Further, the anomeric signal for the bisecting GlcNAc, labelled Gn(III) in figure 4, is in an anomalous position, i.e. it is further upfield than expected, in a position usually occupied by the anomeric hydrogen signal for Gal. This sort of result does not prove three-dimensional distortion by the bisecting GlcNAc but is certainly compatible with such a concept.

The three-dimensional structures of oligosaccharides and their interactions with proteins are relevant not only to the understanding of glycosyltransferase substrate specificity but also to the biological roles that complex carbohydrates play on the surfaces of cells and elsewhere. Our knowledge is currently very preliminary, but special n.m.r. techniques (e.g. the use of nuclear Overhauser enhancement) are currently being used by our group to study this problem.

This research was supported by the Medical Research Council of Canada and by N.I.H. grant no. HD-07889.

REFERENCES

Carver, J. P. & Grey, A. A. 1981 Determination of glycopeptide primary structure by 360 MHz proton magnetic resonance spectroscopy. *Biochemistry, Wash.* **20**, 6607–6616.

Chapman, A., Fujimoto, K. & Kornfeld, S. 1980 The primary glycosylation defect in class E Thy-1-negative mutant mouse lymphoma cells is an inability to synthesize dolichol-P-mannose. *J. biol. Chem.* **255**, 4441–4446.

Cossu, G., Warren, L., Boettiger, D., Holtzer, H. & Pacifici, M. 1982 Similar glycopeptides in normal chondroblasts and in rous sarcoma virus-transformed fibroblasts. *J. biol. Chem.* **257**, 4463–4468.

Gleeson, P., Narasimhan, S., Vella, G. & Schachter, H. 1982*b* Hen oviduct *N*-acetylglucosaminyltransferase IV (Gn-T IV) responsible for branching of complex *N*-glycosyl oligosaccharides. In *Proc. XI International Carbohydrate Symposium, Vancouver, B.C.* (In the press.)

Gleeson, P., Vella, G., Narasimhan, S. & Schachter, H. 1982*a* Hen oviduct *N*-acetylglucosaminyltransferase IV (Gn-T IV) responsible for branching of complex *N*-glycosyl oligosaccharides. *Fedn Proc. Fedn Am. Socs exp. Biol.* **41**, 1147.

Gottlieb, C., Baenziger, J. & Kornfeld, S. 1975 Deficient uridine diphosphate-*N*-acetylglucosamine:glycoprotein *N*-acetylglucosaminyltransferase activity in a clone of Chinese hamster ovary cells with altered surface glycoproteins. *J. biol. Chem.* **250**, 3303–3309.

Gottlieb, C., Skinner, A. M. & Kornfeld, S. 1974 Isolation of a clone of Chinese hamster ovary cells deficient in plant lectin-binding sites. *Proc. natn. Acad. Sci. U.S.A.* **71**, 1078–1082.

Harpaz, N. & Schachter, H. 1980*a* Control of glycoprotein synthesis. Bovine colostrum UDP-*N*-acetylglucosamine: α-D-mannoside β-2-*N*-acetylglycosaminyltransferase I. Separation from UDP-*N*-acetylglucosamine: α-D-mannoside β-2-*N*-acetylglucosaminyltransferase II. Partial purification and substrate specificity. *J. biol. Chem.* **255**, 4885–4893.

Harpaz, N. & Schachter, H. 1980*b* Control of glycoprotein synthesis. Processing of asparagine-linked oligosaccharides by one or more rat liver Golgi α-D-mannosidases dependent on the prior action of UDP-*N*-acetylglucosamine: α-D-mannoside β-2-*N*-acetylglucosaminyltransferase I. *J. biol. Chem.* **155**, 4894–4902.

Johnston, I. R., McGuire, E. J., Jourdian, G. W. & Roseman, S. 1966 Incorporation of *N*-acetyl-D-glucosamine into glycoproteins. *J. biol. Chem.* **241**, 5735–5737.

Johnston, I. R., McGuire, E. J. & Roseman, S. 1973 Sialic acids. XVII. A uridine diphosphate *N*-acetylglucosamine:glycoprotein *N*-acetylglucosaminyltransferase from goat colostrum. *J. biol. Chem.* **248**, 7281–7288.

Kornfeld, S., Gregory, W. & Chapman, A. 1979 Class E Thy-1-negative mouse lymphoma cells utilize an alternate pathway of oligosaccharide processing to synthesize complex-type oligosaccharides. *J. biol. Chem.* **254**, 11 649–11 654.

Li, E. & Kornfeld, S. 1978 Structure of the altered oligosaccharide present in glycoproteins from a clone of Chinese hamster ovary cells deficient in *N*-acetylglucosaminyltransferase activity. *J. biol. Chem.* **253**, 6426–6431.

Liang, C.-J., Yamashita, K., Muellenberg, C. G., Shichi, H. & Kobata, A. 1979 Structure of the carbohydrate moieties of bovine rhodopsin. *J. biol. Chem.* **254**, 6414–6418.

Longmore, G. D. & Schachter, H. 1982 Product identification and substrate specificity studies of the GDP-L-fucose:2-acetamido-2-deoxy-β-D-glucoside(Fuc-Asn-linked GlcNAc) 6-alpha-L-fucosyltransferase in a Golgi-rich fraction from porcine liver. *Carbohydr. Res.* **100**, 365–392.

Munro, J. R., Narasimhan, S., Wetmore, S., Riordan, J. R. & Schachter, H. 1975 Intracellular localization of GDP-L-fucose:glycoprotein and CMP-sialic acid:apolipoprotein glycosyltransferases in rat and pork liver. *Arch. Biochem. Biophys.* **169**, 269–277.

Narasimhan, S., Stanley, P. & Schachter, H. 1977 Control of glycoprotein synthesis. Lectin-resistant mutant containing only one of two distinct *N*-acetylglucosaminyltransferase activities present in wild type Chinese hamster ovary cells. *J. biol. Chem.* **252**, 3926–3933.

Narasimhan, S., Tsai, D. & Schachter, H. 1981 *a* An *N*-acetylglucosaminyltransferase (Gn-T) in hen oviduct which adds GlcNAc to the β-linked Man of the tri-mannosyl core of Asn-linked oligosaccharides. *Fedn Proc. Fedn Am. Socs exp. Biol.* **40**, 1597.

Narasimhan, S., Tsai, D. & Schachter, H. 1981 *b* An *N*-acetylglucosaminyltransferase (Gn-T) in hen oviduct membranes which adds GlcNAc to the β-linked Man of the trimannosyl core of Asn-linked oligosaccharides: *in vitro* synthesis of the 'bisecting' GlcNAc residue. In *Glycoconjugates (Proc. of the Sixth International Symposium on Glycoconjugates)* (ed. T. Yamakawa, T. Osawa & S. Handa), pp. 282–283. Tokyo: Japan Scientific Societies Press.

Ogata, S.-I., Muramatsu, T. & Kobata, A. 1976 New structural characteristic of the large glycopeptides from transformed cells. *Nature, Lond.* **259**, 580–582.

Oppenheimer, C. L., Eckhardt, A. E. & Hill, R. L. 1981 The non-identity of porcine *N*-acetylglucosaminyl-transferases I and II. *J. biol. Chem.* **256**, 11 477–11 482.

Oppenheimer, C. L. & Hill, R. L. 1981 Purification and characterization of a rabbit liver α-1-3-mannoside β-1-2-*N*-acetylglucosaminyltransferase. *J. biol. Chem.* **256**, 799–804.

Rachesky, M. H., Hard, G. C. & Glick, M. C. 1982 Membrane glycopeptides from chemically transformed cells: comparison between mesenchymal and epithelial cell lines derived from dimethylnitrosamine-treated rat kidney. *Cancer Res.* **42**, 39–43.

Rearick, J. I., Chapman, A. & Kornfeld, S. 1981 Glucose starvation alters lipid-linked oligosaccharide biosynthesis in Chinese hamster ovary cells. *J. biol. Chem.* **256**, 6255–6261.

Schachter, H. & Roseman, S. 1980 Mammalian glycosyltransferases: their role in the synthesis and function of complex carbohydrates and glycolipids. In *The biochemistry of glycoproteins and proteoglycans* (ed. W. J. Lennarz), ch. 3, pp. 85–160. New York and London: Plenum Press.

Schachter, H. & Williams, D. 1982 Biosynthesis of mucus glycoproteins. In *Mucus in health and disease* (ed. E. N. Chantler, J. B. Elder & M. Elstein), vol. II, pp. 3–28. New York and London: Plenum Press.

Stanley, P. 1980 Surface carbohydrate alterations of mutant mammalian cells selected for resistance to plant lectins. In *The biochemistry of glycoproteins and proteoglycans* (ed. W. J. Lennarz), ch. 4, pp. 161–189. New York and London: Plenum Press.

Tabas, I. & Kornfeld, S. 1978 The synthesis of complex-type oligosaccharides. III. Identification of an α-D-mannosidase activity involved in a late stage of processing of complex-type oligosaccharides. *J. biol. Chem.* **253**, 7779–7786.

Tabas, I., Schlesinger, S. & Kornfeld, S. 1978 Processing of high mannose oligosaccharides to form complex type oligosaccharides on the newly synthesized polypeptides of the vesicular stomatitis virus G protein and the IgG heavy chain. *J. biol. Chem.* **253**, 716–722.

Takasaki, S., Ikehira, H. & Kobata, A. 1980 Increase of asparagine-linked oligosaccharides with branched outer chains caused by cell transformation. *Biochem. biophys. Res. Commun.* **92**, 735–742.

Tulsiani, D. R. P., Hubbard, S. C., Robbins, P. W. & Touster, O. 1981 Separation and roles of three rat liver Golgi α-mannosidases acting on intermediates in glycoprotein biosynthesis. *Fedn Proc. Fedn Am. Socs exp. Biol.* **40**, 1883.

Tulsiani, D. R. P., Hubbard, S. C., Robbins, P. W. & Touster, O. 1982 Alpha-D-mannosidases of rat liver Golgi membranes. Mannosidase II is the GlcNAcMan(5)-cleaving enzyme in glycoprotein biosynthesis and mannosidases I A and I B are the enzymes converting Man(9) precursors to Man(5) intermediates. *J. biol. Chem.* **257**, 3660–3668.

Turco, S. J., Rush, J. S. & Laine, R. A. 1980 Presence of erythroglycan on human K-562 chronic myelogenous leukemia-derived cells. *J. biol. Chem.* **255**, 3266–3269.

Van den Eijnden, D. H., Joziasse, D. H., Dorland, L., Van Halbeek, H., Vliegenthart, J. F. G. & Schmid, K. 1980 Specificity in the enzymic transfer of sialic acid to the oligosaccharide branches of bi- and triantennary glycopeptides of α-1 acid glycoprotein. *Biochem. biophys. Res. Commun.* **92**, 839–845.

Warren, L., Buck, C. A. & Tuszynski, G. P. 1978 Glycopeptide changes and malignant transformation – a possible role for carbohydrate in malignant behavior. *Biochim. biophys. Acta* **516**, 97–127.

Williams, D., Longmore, G., Matta, K. L. & Schachter, H. 1980 Mucin synthesis. II. Substrate specificity and product identification studies on canine submaxillary gland UDP-N-acetylglucosamine:Gal-β-1-3-GalNAc(GlcNAc-GalNAc) β-6-N-acetylglucosaminyltransferase. *J. biol. Chem.* **255**, 11253–11261.

Williams, D. & Schachter, H. 1980 Mucin synthesis. I. Detection in canine submaxillary glands of an N-acetylglucosaminyltransferase which acts on mucin substrates. *J. biol. Chem.* **255**, 11247–11252.

Phil. Trans. R. Soc. Lond. B **300**, 161–172 (1982)

Printed in Great Britain

Structure and biosynthesis of histocompatibility antigens (H-2, HLA)

By B. Dobberstein, S. Kvist and Lynne Roberts†

European Molecular Biology Laboratory, Postfach 10.2209, D-6900 Heidelberg, F.R.G.

Histocompatibility antigens (H-2K, D and L, and HLA-A, B and C) are highly polymorphic cell surface proteins. Their primary structure has been determined by sequencing the protein, complementary DNAs (cDNAs) or genes in several laboratories. H-2L^d and K^d antigens are encoded by eight separate exons: one encodes the signal sequence, three encode the external domains, one encodes the membrane spanning segment and three encode the cytoplasmic domain. A similar structural organization has been found for an HLA gene.

H-2 and HLA antigens are synthesized on membrane-bound ribosomes and are co-translationally inserted into the membrane of the endoplasmic reticulum. Here they assemble with β_2-microglobulin, a small secretory protein. We describe the structure, the membrane insertion *in vitro* and *in vivo*, the intracellular transport and the surface expression of these antigens.

INTRODUCTION

Glycoproteins of the plasma membrane play important roles in several aspects of cellular recognition (see Warren 1980). Among these proteins are the major histocompatibility antigens called H-2K, D and L in mouse and HLA-A, B and C in man (Klein 1979; Ploegh *et al.* 1981). As they were originally detected as the targets for aggressive T cells in the graft rejection they are also called the major transplantation antigens. Their physiological role might, however, be their function as a restricting element in T-cell killing of infected self cells (Zinkernagel & Doherty 1974).

In recent years progress has been made in gaining an understanding of how plasma membrane proteins are synthesized, are inserted into the membrane, become modified and finally appear on the cell surface. Much has been learned by studying the biogenesis of viral membrane proteins (Katz *et al.* 1977; Garoff *et al.* 1978, 1982; Tabas & Kornfeld 1978). Because they are made in large quantities in the infected cell they are easily approachable for biochemical analysis. The H-2 and HLA antigens became a useful system with the availability of antibodies of high specificity and titre. In this way, the obstacle posed by the low cellular content of H-2 and HLA antigens could be overcome, allowing studies on the biosynthesis and cell-surface expression of these antigens. They are particularly interesting as they are composed of two subunits: one membrane-integrated and one soluble. The primary structure of an H-2 and HLA antigen has been determined recently and genes coding for them have been characterized (Orr *et al.* 1979; Trägårdh *et al.* 1980; Nathenson *et al.* 1981; Moore *et al.* 1982; Evans *et al.* 1982; Kvist *et al.* 1982; Malissen *et al.* 1982). Here we describe the structure and the stages in the maturation of H-2 and HLA antigens, established by work in several laboratories.

† Present address: Department of Biological Sciences, University of Warwick, Coventry CV4 7AL, U.K.

STRUCTURAL FEATURES

(a) The extracellular portion

On the cell surface H-2 and HLA antigens are composed of a polymorphic heavy chain (43–48 kDa) and a non-covalently linked small polypeptide, β_2-microglobulin (11.5 kDa) (Cresswell *et al.* 1974; Peterson *et al.* 1974). The heavy chain is glycosylated and spans the membrane (Parham *et al.* 1977; Nathenson & Cullen 1974; Walsh & Crumpton 1977). The amino terminus of the molecule is exposed extracellularly and this portion can be cleaved close to the membrane by papain (see figure 1) (Coligan *et al.* 1981; Henning *et al.* 1976).

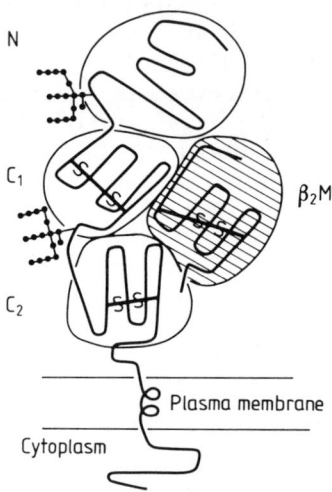

FIGURE 1. Schematic representation of a histocompatibility antigen class 1 H-2 molecule, modified after Coligan *et al.* (1981). The domains shown are indicated as follows: N, N-terminal domain; C_1 and C_2, cystine-containing domains; β_2M, β_2-microglobulin polypeptide.

The primary structure of H-2 and HLA antigens has been determined either by sequence analysis of the protein itself or by sequencing cDNAs or genes coding for them (Coligan *et al.* 1981; Lalanne *et al.* 1982; Moore *et al.* 1982; Ploegh *et al.* 1981*a*; Kvist *et al.* 1982; Steinmetz *et al.* 1981; Malissen *et al.* 1982). The H-2K^b antigen (an H-2 antigen coded by a gene of the K locus of the b allele) is 346 amino acids long; the HLA-B7 antigen (an HLA antigen coded by a gene in the B locus of the allele 7) is 337 amino acids long (Coligan *et al.* 1981; Ploegh *et al.* 1981*a*). A cluster of hydrophobic or uncharged amino acids is located close to the carboxy-terminus between amino acids 283 and 307 (Nathenson *et al.* 1981). This segment most probably spans the membrane. The cysteine residues forming intrachain disulphide linkages are found at residues 101, 164, 203 and 259 in the H-2K^b molecule (Coligan *et al.* 1981). Based on suscepti-bility to cleavage by acid or proteolytic enzymes, as well as the location of the two intrachain disulphide linkages, the extracellular part can be divided into three domains. They have been designated N (α1) for the amino-terminal domain, C1 (α2) and C2 (α3) for the cysteine-con-taining domains (Terhorst *et al.* 1977; Nathenson *et al.* 1981). Such an organization is also supported by the location of the exon-intron boundaries in the genes coding for H-2 and HLA antigens; three exons encode the extracellular portion and the size of each of them is in excellent agreement with the size of the domains suggested by the analysis of the protein itself

(Moore *et al.* 1982; Malissen *et al.* 1982; Kvist *et al.* 1982). The N-terminal domain is comprised of residues 1–90, the C1 domain 91–182 and the C2 domain 183–274 (see figure 2) (Moore *et al.* 1982; Kvist *et al.* 1982; Malissen *et al.* 1982).

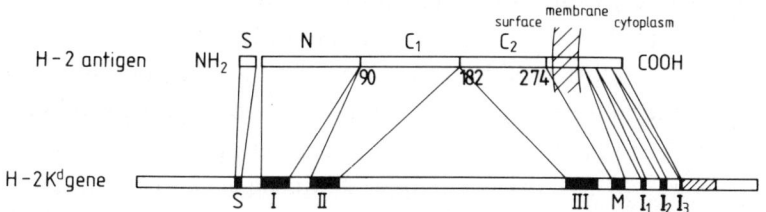

FIGURE 2. Organization of the *H-2K^d* gene. Comparison of its exon–intron structure with the H-2K^b antigen and the cDNA for an H-2K^d antigen (Coligan *et al.* 1980; Kvist *et al.* 1982). Exons corresponding to protein domains are shown as filled boxes, introns as open boxes, and the 3′ untranslated region as a hatched box. The exons are indicated by: S, coding for the signal sequence; I, II, III, coding for the external protein domains N, C₁ and C₂; M, coding for the membrane domains; and I₁, I₂, I₃, coding for the intracellular segment.

```
                 SURFACE              MEMBRANE            CYTOPLASM

        H-2 antigens                                                      references
                                      _____
                                                              * ***
        H-2Kᵇ            EPPSTVSNMATVAVLVVLGAAIVTGAVVAFVMKMRRRAT              1

pH-2ᵈ-4, λH-2Kᵈ  ▲L-------TVII--------------------- --N-▲                    2,3

      pH-2ᵈ-1     P-S--KT-TVII--P-----VVIL---M----- ---N-                     4

pH-2ᵈ-3, λH-2Lᵈ  ▲P---F-D-Y-VI----G----MAII--------- ---N-▲                   5,6

     λ27.1(Qa?)  ▲P--Y-------I--V-D---VAII--------N --Trm                     7

        HLA-antigens

        HLA-B7          LPSSQSTVP·VG·VAG·AV·AVVV·GAVVAAVMCRRKSS               8

      λHLA-12    ▲E----P---I--I---LVLLVA--T--------W-K--▲                     9
```

FIGURE 3. Comparison of the amino acid sequence of the membrane-spanning segment. The sequence is expressed in the one-letter code. Dashes indicate homology to the H-2K^b and HLA-B7 sequence respectively. Dots indicate that the amino acid at that site is not known. The sequence has been obtained by sequencing the protein (H-2K^b, HLA-B7), cDNA (pH-2^d-4, pH-2^d-1, pH-2^d-3) or the gene (λ27.1, λHLA-12, λH-2L^d, λH-2K^d). The arrow indicates a splice site in the gene. Basic amino acid residues at the cytoplasmic side are indicated by an asterisk. References: 1, Coligan *et al.* (1981); 2, Lalanne *et al.* (1982); 3, Kvist *et al.* (1982); 4, Kvist *et al.* (1981); 5, Bregegere *et al.* (1981); 6, Moore *et al.* (1982); 7, Steinmetz *et al.* (1981); 8, Ploegh *et al.* (1981); 9, Malissen *et al.* (1982).

The number of carbohydrate units on H-2 and HLA antigens can vary greatly. Only one unit is found in HLA antigens, whereas up to three are present in H-2 antigens. They are linked to an asparagine at position 86, in H-2 antigens also on asparagine 176, and in the H-2K^d molecule on residue 256 as well (Parham *et al.* 1977; Nathenson *et al.* 1981; Kvist *et al.* 1982).

(b) The membrane-spanning segment

By using proteases as a dissecting tool, the location of the membrane-spanning segment of H-2 and HLA antigens has been determined (see below) (Dobberstein *et al.* 1979; Owen *et al.* 1980). It is made up of about 30 amino acids and it located about 30 residues away from the carboxy terminus. One would predict that such a segment contains a stretch of uncharged or hydrophobic residues, or both, that are able to interact with the hydrophobic lipid bilayer. A segment with this property is contained in the amino acid sequence of H-2 and HLA antigens. Figure 3 shows a comparison of the amino acid sequences of the membrane-spanning segments. Although the amino acids may differ, the hydrophobic character of this segment was conserved in all chains sequenced so far. Changes were detected mainly in the region close to the external side of the membrane. The cluster of basic amino acid residues on the cytoplasmic side is conserved, suggesting a general function of this region in anchoring the protein in the membrane.

A separate exon codes for the membrane spanning segment plus some amino acid residues on either side (see figures 2 and 3) (Moore *et al.* 1982; Malissen *et al.* 1982; Kvist *et al.* 1982). This arrangement leads one to ask if a functional membrane segment is composed of three portions: (1) amino acid residues forming a link between the membrane and external domains, (2) the lipid bilayer-spanning segment, and (3) the cluster of basic amino acid residues on the cytoplasmic side of the membrane.

(c) *The cytoplasmic segment*

Little is known about the function of the cytoplasmic segment of plasma membrane proteins. It has been suggested that their function is to transmit 'signals' from the outside to the inside of the cell by interacting with cytoplasmic constituents (Bourguignon & Singer 1977; Koch & Smith 1978; Pober *et al.* 1981). No such functional interaction with cytoplasmic elements has yet been demonstrated. The size of the cytoplasmic segment can be estimated on antigens that are inserted into microsomes. Microsomes are closed vesicles derived from the endoplasmic reticulum, where the insertion of newly synthesized plasma membrane proteins like H-2 and HLA antigens occurs (see below). When these vesicles were treated with protease only the portion of the antigen that protruded from the membrane into the cytoplasm was digested. The remainder of the molecule was protected against digestion by the permeability barrier of the membrane. After separation of the intact and proteolysed antigens on sodium dodecyl sulphate (SDS)–polyacrylamide gels, the size of the cytoplasmic segment could be calculated from their difference in molecular mass. The size of the cytoplasmic segment of H-2 and HLA antigens has been estimated to contain about 30 amino acid residues (Dobberstein *et al.* 1979; Owen *et al.* 1980). This estimated size is in good agreement with the 30–40 amino acid residues determined by sequence analysis of H-2 and HLA proteins, cDNAs and genes. Figure 4 shows sequences of cytoplasmic segments. Here the sequences are most varied, in particular with regard to their size. The strong variation may either indicate the existence of little functional constraint on this segment or reflect different modes of interaction with cytoplasmic components.

The cytoplasmic segment is encoded in H-2 genes by three exons and in an HLA gene by two (Moore *et al.* 1982; Kvist *et al.* 1982; Malissen *et al.* 1982). As the exons encoding signal sequences, cytoplasmic and membrane domains all correlate very well with functionally or structurally defined regions, it is tempting to speculate that separate functions might also be indicated by the different exons coding for the cytoplasmic segments. Such a hypothesis can be tested. H-2 genes, isolated by molecular cloning, can be expressed after stable transformation in mouse L cells (Goodenow *et al.* 1982; Burgert *et al.* 1982; Evans *et al.* 1982). H-2 antigens became expressed and can function on the cell surface as restricting elements in T-cell killing (Örn *et al.* 1982). This system should also allow one to test modified genes, e.g. those modified in their cytoplasmic segment. H-2 genes could be expressed that are modified with regard to the number and composition of the exons coding for the cytoplasmic segment. It would for instance be important to establish whether a particular C-terminal segment is required for the T-cell killing of an infected cell.

(*d*) β_2-*Microglobulin*

The small subunit of H-2 and HLA antigens has been identified as β_2-microglobulin (Cresswell *et al.* 1974; Peterson *et al.* 1974; Rask *et al.* 1974). It has a molecular mass of about 12 kDa and contains 99 and 100 amino acid residues in mouse and man respectively (Cunningham *et al.* 1973; Gates *et al.* 1981). Based on the homology between immunoglobulin constant regions and

```
                              CYTOPLASMIC SEGMENTS
H-2 antigens                                                             references
                        * ***
H-2K^b                 MKMRRRAT GGKGGDYALAP GSQTSDLSLPDCK VMVHPPHS              1
pH-2^d-4, λH-2K^d      --- --N- ----VN----- -Y---------G- ----D---LA            2,3
                               ▲           ▲             ▲
pH-2^d-1               -- ---N- ----------- ---S--M------ -                     4
pH-2II                 -- ---N- ----------- ---S-EM--R--- GDTLGSDWGGAMWT        7
pH-2^d-3, λH-2L^d      -- ---N- ----------- ---S-EM--R-- KA                     5,6
                               ▲           ▲            ▲
HLA antigens
HLA-B7                 MCRRKSS GGKGGSYSQAA CSDSAQGSDVSLTA                       10
HLA-A2                 -W----- DR--------- S-------------                       10
λHLA-12                -W-K--- DR--------- S-N-----------                       9
                              ▲           ▲
```

FIGURE 4. Comparison of cytoplasmic segments. Sequences have been determined by sequencing the protein (H-2K^b, HLA-B7, HLA-A2), cDNA (pH-2^d-4, pH-2^d-1, pH 2II, pH-2^d-3) or the gene (λH-2L^d, λH-2K^d, λHLA-12). Dashes indicate sequence homology with the H-2K^b and HLA-B7 antigen respectively. Arrow indicate a splice site in the gene. The cluster of basic amino acids is indicated by asterisks. References as in figure 3, plus: 10, Robb *et al.* (1978).

the C2 domain of H-2 antigens, it has been suggested that it interacts with the C2 domain (Ploegh *et al.* 1981*a*). This is supported by the finding that the exons encoding the C2 domain are highly conserved between different H-2 genes (Steinmetz *et al.* 1982). This conservation could reflect the evolutionary constraint on this domain posed by the interaction with β_2-microglobulin. The function of β_2-microglobulin in the oligomeric complex with the heavy chains is unknown.

BIOSYNTHESIS

Early studies on the biosynthesis of H-2 and HLA antigens indicated that they were synthesized at an intracellular site and were subsequently transported to the plasma membrane (Vitetta & Uhr 1975). It was suggested that they are synthesized and transported like secretory proteins, going from the rough endoplasmic reticulum via the Golgi complex to the plasma membrane. Pulse–chase techniques and cell fractionation were then used to follow the flow of newly synthesized membrane constituents to the cell surface (Krangel *et al.* 1979; Dobberstein *et al.* 1979; Owen *et al.* 1980; Croze & Morré 1981; Tartakoff *et al.* 1981). Labelled H-2 antigens appeared first in fractions containing membranes derived from the endoplasmic reticulum, then from the Golgi apparatus and last from the plasma membrane.

(a) Membrane insertion

The heavy chain and β_2-microglobulin both have to cross a membrane – at least in part for the heavy chain – to go from their site of synthesis to their final location, the plasma membrane. Steps in the biosynthesis of plasma membrane proteins can be investigated in cell-free protein synthesizing systems as well as *in vivo* by metabolic labelling. Both of these approaches have been used for studying H-2 and HLA antigens.

Heavy chains and β_2-microglobulin are translated from separate mRNAs. These can be separated on a sucrose gradient and show sedimentation values of 17*S* and 9*S*, respectively (Dobberstein *et al.* 1979; Jay *et al.* 1979).

```
                        SIGNAL SEQUENCE        MATURE
                                               PROTEIN
       pre-H-2 antigens
                                                                references
                        **+ +*********+    +*+
       λH-2Kᵈ           MAPCTLLLLLAAALAPTQTRA GPHSLRYFVT           3
       H-2Kᵈ                                  G·HSLRYFVT          11
       λH-2Lᵈ           MAPRTLLLLLAAAWPDSDPR  GPHSMRYFET          6,12
       H-2Lᵈ                                  ··H·MRYF·T          11
       λ27.1(Qa?)       MALTMLLLLVAAALTLIETRA GQHSLQYFHT           7
       pre-HLA antigens
       λHLA-12          MAPRTLLLLLSGALALTQTWA RSHSMRYFYT           9
       HLA-A,B,C                              GSHSMRYFYT          13
       pre-ß₂-microglobulin
 mouse pre-ß₂-M         S·SV·LVFLVLVSL·GLY·  ····P···VY          14
       ß₂-M                                   IQKTPQIQVY          15
 human pre-ß₂-M             ...LALLSLSGLQA    IQRTPKIQVY          16
       ß₂-M                                   IQRTPKIQVY          17
```

FIGURE 5. Comparison of signal sequences. Sequences were determined by sequencing proteins (H-2Kᵇ, H-2Lᵈ, β_2-microglobulin), cDNA (pre-β_2-microglobulin) or genes (λH-2Kᵈ, λH-2Lᵈ, λ27.1). Dots indicate an unknown amino acid at that position. Arrows point at the site of postulated signal peptidase cleavage and splice site in the gene. An asterisk indicates an amino acid residue conserved in signal sequences of all three pre-H-2 antigens, a cross indicates an amino acid conserved in two. References as in figure 3, plus: 11, Coligan *et al.* (1980); 12, Evans *et al.* (1982); 13, Trägårdh *et al.* (1980); 14, Lingappa *et al.* (1979); 15, Gates *et al.* (1981); 16, Suggs *et al.* (1981); 17, Cunningham *et al.* (1973).

When mRNA from cells expressing large amounts of H-2 and HLA antigens were translated in a cell-free system, both were synthesized as higher molecular mass precursors (Dobberstein *et al.* 1979; Ploegh *et al.* 1979). The heavy chain (pre-H-2Dᵈ) was found to be 2 kDa heavier than its non-glycosylated, membrane-inserted counterpart (see below). Precursors to a large number of secretory and some membrane proteins have been characterized (Kreil 1981). At their amino termini they contain a signal sequence that is cleaved from the nascent chain during its insertion into the membrane of the endoplasmic reticulum. It is thought to direct the nascent polypeptide chain to the endoplasmic reticulum and facilitate its insertion into the membrane (see Blobel 1980). Figure 5 shows the signal sequences of some H-2 and HLA antigens (Moore *et al.* 1982; Kvist *et al.* 1982; Malissen *et al.* 1982). They were deduced from the nucleotide sequences of their genes. The signal sequences in these genes are encoded by separate exons (see figure 5). They comprise 20 or 21 amino acid residues and contain a central region devoid of charged residues, the characteristic feature of a signal sequence. The NH_2-terminal half of the signal sequence of H-2 antigens shows a strikingly higher degree of conservation than the region close to the site of signal peptidase cleavage (see figure 5). This could indicate that higher constraint is exerted on the amino-terminal half. The stretch of hydrophobic amino acids is located in this portion, which would strongly support the notion that such a hydrophobic region is essential for signal sequence function. Amino acids with short side chains, such as Gly, Ala, Cys and Ser, are usually present at the site of signal sequence cleavage (Kreil 1981). In the H-2L^d antigen an asparagine residue has been found at this site. A sequencing error is unlikely as two groups obtained the same result independently (Moore *et al.* 1982; Evans *et al.* 1982). Because the sequence was deduced from a genomic clone it would be important to confirm the splice site by sequencing part of the cDNA or the preprotein itself.

As already mentioned, the insertion of proteins into a membrane can be reproduced *in vitro* by using rough microsomes derived from dog pancreas (Blobel & Dobberstein 1975). In such a system H-2 antigens were inserted co-translationally into the membrane (Dobberstein *et al.* 1979). The concomitant biosynthesis of β_2-microglobulin was not required for the insertion of the heavy chains into the membrane. The disposition of the heavy chain in the membrane of the endoplasmic reticulum was the same as that on the cell surface, i.e. the carboxy terminus was on the cytoplasmic side and the amino terminus on the luminal side.

Heavy chains inserted into microsomal membranes *in vitro* became glycosylated. This was evidenced by a shift to a higher molecular mass when these antigens were separated on an SDS–polyacrylamide gel, and by their binding to lentil lectin (Dobberstein *et al.* 1979).

Heavy chains labelled for 5 min *in vivo* were also glycosylated because they also bound to lentil lectin. Their carbohydrate unit was sensitive to endoglycosidase H (Owen *et al.* 1980; Sege *et al.* 1981; Krangel *et al.* 1979). This enzyme is known to cleave polymannose-type carbohydrates linked to asparagine (Tarentino & Maley 1974).

The glycosylation of the heavy chains was blocked *in vivo* by tunicamycin, an inhibitor of *N*-linked glycosylation (Ploegh *et al.* 1981 *b*). Glycosylation was required neither for the insertion of heavy chains into the membrane nor for their assembly with β_2-microglobulin. Unglycosylated heavy chains became correctly integrated into the membrane of the endoplasmic reticulum, as evidenced by protection against protease (see above). Antibodies recognizing exclusively the non-glycosylated heavy chain did co-precipitate β_2-microglobulin.

Biosynthesis of β_2-microglobulin proceeds very similarly to that of other secretory proteins (Dobberstein *et al.* 1979; Lingappa *et al.* 1979; Algranati *et al.* 1980). This molecule can in fact be considered as a secretory protein that becomes a peripheral membrane protein only upon association with the membrane integrated heavy chain. In a cell-free system devoid of membranes it was synthesized as a higher molecular mass precursor (pre-β_2-microglobulin) containing an amino-terminal signal sequence (Dobberstein *et al.* 1979; Algranati *et al.* 1980). Figure 5 shows the partial sequences of the signal sequences of β_2-microglobulin from mouse and man (Lingappa *et al.* 1979; Suggs *et al.* 1981). In mouse it comprises 19 amino acids. When β_2-microglobulin was synthesized *in vitro* in the presence of dog pancreas microsomal membranes, it was co-translationally translocated across this membrane and the signal sequence was cleaved. When protease was used to probe the topology of β_2-microglobulin in microsomal vesicles, it was found to be fully protected by the membrane (Dobberstein *et al.* 1979). Thus, in contrast to the heavy chain, β_2-microglobulin is completely translocated across the membrane. This was to be expected because β_2-microglobulin does not contain a stretch of hydrophobic or uncharged amino acid residues that could anchor it in the membrane.

A number of secretory and membrane proteins compete for specific sites on microsomal membranes that are required for their membrane insertion. Proteins required for membrane translocation have been characterized recently. They comprise a 'signal recognition protein' complex and a membrane-integrated receptor, the 'docking protein' (Walter & Blobel 1981; Meyer *et al.* 1982). Both of these proteins function sequentially in the recognition of the signal sequence on the nascent secretory or membrane protein and the subsequent specific interaction with the membrane. That the same mechanism also functions in the translocation of nascent β_2-microglobulin across the membrane is suggested by its competition with ovalbumin for sites in the membranes (Lingappa *et al.* 1979).

(b) Oligomeric assembly

Heavy chains and β_2-microglobulin are independently inserted into or translocated across the membrane of the endoplasmic reticulum. By using antibodies that detect heavy chains only, oligomeric assembly with β_2-microglobulin has been studied (Krangel *et al.* 1979; Owen *et al.* 1980; Dobberstein *et al.* 1979). In such a case β_2-microglobulin recovered in immunoprecipitates must have assembled with the heavy chains. In all cell lines studied, oligomeric assembly occurred very rapidly after newly synthesized heavy chains were inserted into the membrane of the endoplasmic reticulum. After as little as 5 min of pulse labelling of either the human lymphoblastoid cell lines Bri8 or T5-1 or the mouse cell line SL2, β_2-microglobulin was found to be bound to the heavy chains (Krangel *et al.* 1979; Dobberstein *et al.* 1979; Owen *et al.* 1980). Whether nascent heavy chains do already assemble with β_2-microglobulin is unclear. The observed kinetics of association found in the cell line T5-1 suggests that β_2-microglobulin *can* assemble with completed heavy chains. β_2-Microglobulin is synthesized in excess over the heavy chains, at least twofold in Bri8 cells (Owen *et al.* 1980). Excess β_2-microglobulin does not appear to be associated with other polypeptide chains. Results from pulse–chase experiments in Bri8 and SL2 cells are consistent with synthesis of an excess of β_2-microglobulin. The amount of labelled β_2-microglobulin increased steadily during the chase period, whereas the intensity of labelled heavy chains remained relatively constant. β_2-Microglobulin that was not associated with heavy chains remained largely in the endoplasmic reticulum (e.r.), whereas that assembled with the heavy chain underwent intracellular transport (Dobberstein *et al.* 1979). The fate of excess β_2-microglobulin in the e.r. has not yet been determined. It may be eventually secreted; however, this would proceed more slowly than the assembled antigens appear on the cell surface. Excess free β_2-microglobulin in the e.r. would provide the condition whereby newly synthesized heavy chains would assemble with an existing pool of β_2-microglobulin.

INTRACELLULAR TRANSPORT

It is not known which determinants regulate the intracellular transport of plasma membrane proteins from the e.r. to the cell surface. From studies with viral membrane proteins it appears that these determinants can be different from one protein to another. Among the modifications that have been suggested to function in intracellular transport are glycosylation, phosphoryla-tion, fatty acid acylation and oligomeric assembly. It is essential to determine how these modifica-tions affect the surface expression of H-2 or HLA antigens. A schematic illustration of the intracellular pathway of H-2 and HLA antigens is shown in figure 6.

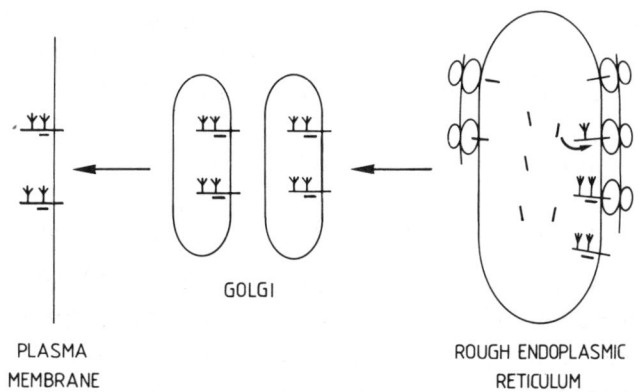

FIGURE 6. Schematic illustration of the intracellular pathway of H-2 and HLA antigens. Y, Oligosaccharide; −, β₂-microglobulin, ϤϤ, H-2 and HLA heavy chain.

(a) Carbohydrate modifications

The carbohydrate portions of H-2 and HLA antigens are modified during intracellular trans-port in a manner similar to that of viral glycoproteins (Krangel et al. 1979; Owen et al. 1980; Ploegh et al. 1981; Sege et al. 1981). Asparagine-linked carbohydrate structures and their processing during intracellular transport are well characterized for a number of viral membrane proteins, in particular for the VSV-G protein (Tabas & Kornfeld 1978). Polymannose-type oligosaccharide chains are transferred in the rough endoplasmic reticulum onto the nascent polypeptide chain (Rothman & Lodish 1977; Kruppa 1979). After the removal of glucose and mannose residues, the mature complex-type oligosaccharide arises by the addition of N-acetyl glycosamine, galactose, fucose and sialic acid residues (Tabas et al. 1978; Tabas & Kornfeld 1978). While the polymannose-type oligosaccharide is sensitive to endoglycosidase H (endo H), the complex-type is not.

As judged by SDS–polyacrylamide gel electrophoresis, the apparent size of H-2K^d, D^d and HLA antigens increases 20–30 min after their synthesis (Dobberstein et al. 1979; Krangel et al. 1979; Sege et al. 1981). This usually indicates the conversion of the polymannose-type to complex-type carbohydrate. Likewise the endo H sensitive form of HLA antigens is modified to an insensitive one. Neuraminidase had no detectable effect on the endo H sensitive form; however, the endo H insensitive form was converted to a form of lower molecular mass (Sege et al. 1981). This is consistent with the modification of a polymannose-type carbohydrate unit to a complex oligosaccharide unit containing sialic acid residues.

The carbohydrate portion is not necessary for the surface appearance of HLA antigens (Owen et al. 1980; Ploegh et al. 1981). Bri8 and JY cells treated with tunicamycin expressed HLA antigens on their surface. This was demonstrated by binding antibodies to intact cells before lysis and immunoprecipitation. Also, glycosylated and non-glycosylated heavy chains appeared on the cell surface at the same time, about 30 min after synthesis in both cell types studied.

The Golgi complex is the intracellular location where conversion of polymannose-type to complex-type oligosaccharide is thought to occur. Passage of H-2^d antigens through the Golgi complex has been investigated by Tartakoff *et al.* (1981) using the ionophone monensin. This drug has been shown to interrupt the intracellular transport of secretory and viral membrane proteins at the level of the Golgi complex and to cause their accumulation within dilated Golgi cisternae. The proteins remain incompletely glycosylated, lacking fucose, galactose and sialic acid. Terminal glycosylation of H-2^d antigens was blocked by monensin. Furthermore, indications were obtained that H-2^d antigens in monensin-treated cells do not reach the cell surface. Thus H-2 antigens normally pass through the Golgi subsite defined by monensin and acquire terminal sialic acid distal to this site (Tartakoff *et al.* 1981).

(b) Assembly of HLA antigens with β$_2$-microglobulin

From studies of HLA antigens in the human lymphoblastoid cell line Daudi it has been concluded that β$_2$-microglobulin is required for the cell-surface expression of the heavy chains (Ploegh *et al.* 1979; Owen *et al.* 1980; Sege *et al.* 1981; Goodfellow *et al.* 1975). This cell line does not synthesize β$_2$-microglobulin and does not express the heavy chains on its cell surface. Heavy chains are, however, synthesized by these cells. This was demonstrated convincingly by precipitating HLA heavy chains from pulse-labelled Daudi cells. Furthermore, the precursor of the heavy chains was isolated from the cell-free translation products of Daudi cell mRNA. No β$_2$-microglobulin could be detected in Daudi cells, neither after labelling *in vivo* nor after translation of their mRNA.

The heavy chains became glycosylated and this could be blocked by the inhibitor of *N*-linked glycosylation, tunicamycin. The carbohydrate portion of pulse-labelled heavy chains of Daudi cells remained endo H sensitive throughout a chase period of 90 min (Owen *et al.* 1980). Neuraminidase was unable to modify the chased heavy chains. No fucose was incorporated into heavy chains synthesized by Daudi cells in contrast to those of Bri8 cells (Sege *et al.* 1981). The stability of heavy chains, at least up to 90 min, was not drastically reduced by the lack of association with β$_2$-microglobulin (Owen *et al.* 1980; Sege *et al.* 1981). Taken together these data demonstrate that the heavy chains of Daudi cells are arrested at an early stage of intracellular transport, most probably in the endoplasmic reticulum. They furthermore suggest that the association with β$_2$-microglobulin is necessary for intracellular transport of the heavy chains to the Golgi complex and the cell surface.

The requirement of β$_2$-microglobulin for the intracellular transport of HLA heavy chains is also suggested by studies on a structurally altered HLA-A2 heavy chain (Krangel *et al.* 1982). This chain became glycosylated and thus presumably inserted into the membrane of the endoplasmic reticulum *in vivo*. However, unlike the normal HLA-A2 heavy chains, the variant did not associate with β$_2$-microglobulin; neither did it undergo processing of its polymannose oligosaccharide nor could it be detected on the cell surface. The stability within the cell did not seem to be altered. Krangel *et al.* (1982) suggest that the primary defect in this HLA-A2 variant is the failure to associate stably with β$_2$-microglobulin.

It is not known why heavy chains alone cannot undergo intracellular transport. One possibility is that they aggregate with constitutive elements of the rough endoplasmic reticulum and thereby become unable to accumulate at sites where postulated transport vesicles bud from the endoplasmic reticulum membrane (Warren 1980; Pearse & Bretscher 1981).

(c) Other modifications

Besides glycosylation and assembly with β_2-microglobulin, H-2 and HLA heavy chains become phosphorylated (Rothbard *et al.* 1980; Pober *et al.* 1978). This phosphorylation occurs on a serine residue in the cytoplasmic C-terminal end. The possibility, however, that the phosphorylation of the heavy chains occurs after lysis of the cells with detergent has not yet been ruled out. A second modification, a transglutaminase-catalysed amidation of the intracellular carboxy-terminal segment of HLA-B7 and A2 antigens, has been suggested (Pober & Strominger 1981). Transglutaminase couples amines specifically to glutamine residues. Pober & Strominger (1981) suggest that cellular transglutaminases could be involved in membrane events by catalysing the cross-linking of the cytoplasmic portion of a transmembrane protein to other intracellular constituents. No such labelling *in vivo* has yet been demonstrated. As mentioned above, the use of modified heavy chains will be a means by which questions on the physiological function of these possible modifications could be answered in the future.

With the ability to isolate and express genes coding for plasma membrane proteins, their detailed characterization has become possible. The hydrophobic segments, which posed serious problems for the protein chemist, can be easily analysed by DNA sequencing technology. The expression of modified genes in different host cells will greatly facilitate an evaluation of the functionally relevant segments of the histocompatibility antigens.

The authors are very grateful to Wendy Moses and Annie Steiner for preparing the manuscript, Petra Riedinger for graphics and David Meyer for critical comments and reading of the manuscript.

REFERENCES

Algranati, J. D., Milstein, C. & Ziegler, A. 1980 *Eur. J. Biochem.* **103**, 197–207.
Blobel, G. 1980 *Proc. natn. Acad. Sci. U.S.A.* **77**, 1496–1500.
Blobel, G. & Dobberstein, B. 1975 *J. Cell Biol.* **67**, 852–862.
Bourguignon, L. Y. W. & Singer, S. J. 1977 *Proc. natn. Acad. Sci. U.S.A.* **74**, 5031–5035.
Bregegere, F., Abastado, J. P., Kvist, S., Rask, L., Lalanne, J. L., Garoff, H., Cami, B., Wiman, K., Larhammar, D., Peterson, P. A., Gachelin, G., Kourilsky, P. & Dobberstein, B. 1981 *Nature, Lond.* **292**, 78–81.
Burgert, H. G., Soderberg, K., Rüsch, E., Hämmerling, G. J., Kvist, S. & Dobberstein, B. 1982 *EMBO Jl.* (Submitted.)
Coligan, J. E., Kindt, T. J., Nairn, R., Nathenson, S. G., Sachs, D. H. & Hansen, T. H. 1980 *Proc. natn. Acad. Sci. U.S.A.* **77**, 1134–1138.
Coligan, J. E., Kindt, T. J., Uehara, H., Martinko, J. & Nathenson, S. G. 1981 *Nature, Lond.* **291**, 35–39
Cresswell, P., Springer, T., Strominger, J. L., Turner, M. J., Grey, H. M. & Kubo, R. T. 1974 *Proc. natn. Acad. Sci. U.S.A.* **71**, 2123–2127.
Croze, E. M. & Morré, D. J. 1981 *Proc. natn. Acad. Sci. U.S.A.* **78**, 1547–1551.
Cunningham, B. A., Wang, J. L., Berggard, J. & Peterson, P. A. 1973 *Biochemistry, Wash.* **12**, 4811–4821.
Dobberstein, B., Garoff, H., Warren, G. & Robinson, P. 1979 *Cell* **17**, 759–769.
Evans, G. A., Margulies, D. H., Camerini-Otero, R. D., Ozato, K. & Seidman, J. G. 1982 *Proc. natn. Acad. Sci. U.S.A.* **79**, 1994–1998.
Garoff, H., Kondor-Koch, C. & Riedel, H. 1982 In *Current topics in microbiology and immunology*, vol. 99 (ed. M. Cooper *et al.*), pp. 1–50. Berlin and Heidelberg: Springer-Verlag.
Garoff, H., Simons, K., Dobberstein, B. 1978 *J. molec. Biol.* **124**, 587–600.
Gates, F. T., Coligan, J. E. & Kindt, T. J. 1981 *Proc. natn. Acad. Sci. U.S.A.* **78**, 554–558.
Goodenow, R. S., McMillan, M., Örn, A., Nicholsen, M., Davidson, N., Frelinger, J. A. & Hood, L. 1982 *Science, Wash.* **215**, 677–679.
Goodfellow, P. N., Jones, E. A., Van Heyningen, V., Salomon, E., Bobrow, M., Miggiano, V. & Bodmer, W. F. 1975 *Nature, Lond.* **254**, 267–269.
Henning, R., Milner, R. J., Reske, K., Cunningham, B. A. & Edelman, G. M. 1976 *Proc. natn. Acad. Sci. U.S.A.* **73**, 118–122.

Jay, G., Ferrini, U., Robinson, E. A., Khoury, G. & Apella, E. 1979 *Proc. natn. Acad. Sci. U.S.A.* **76**, 6562–6566.
Katz, F. N., Rothman, J. E., Knipe, D. M. & Lodish, H. F. 1977 *J. supramolec. Struct.* **7**, 353–370.
Klein, J. 1979 *Science, Wash.* **203**, 516–521.
Koch, G. L. E. & Smith, M. J. 1978 *Nature, Lond.* **273**, 274–278.
Krangel, M. S., Orr, H. T. & Strominger, J. L. 1979 *Cell* **18**, 979–991.
Krangel, M. S., Pious, D. & Strominger, J. L. 1982 *J. biol. Chem.* **257**, 5296–5305.
Kreil, G. 1981 *A. Rev. Biochem.* **50**, 317–348.
Kruppa, J. 1979 *Biochem. J.* **181**, 295–300.
Kvist, S., Bregegere, F., Rask, L., Cami, B., Garoff, H., Daniel, F., Wiman, K., Larhammar, D., Abastado, J. P., Gachelin, G., Peterson, P. A., Dobberstein, B. & Kourilsky, P. 1981 *Proc. natn. Acad. Sci. U.S.A.* **78**, 2772–2776.
Kvist, S., Roberts, L. & Dobberstein, B. 1982 *EMBO Jl.* (Submitted.)
Lalanne, J. L., Bregegere, F., Delabre, C., Abastado, J. P., Gachelin, G. & Kourilsky, P. 1982 *Nucl. Acids Res.* **10**, 1039–1044.
Lingappa, V. R., Cunningham, B. A., Jazwinski, S. M., Hopp, T. P., Blobel, G. & Edelman, G. M. 1979 *Proc. natn. Acad. Sci. U.S.A.* **76**, 3651–3655.
Malissen, M., Malissen, B. & Jordan, B. R. 1982 *Proc. natn. Acad. Sci. U.S.A.* **79**, 893–897.
Meyer, D. J., Krause, E. & Dobberstein, B. 1982 *Nature, Lond.* **297**, 647–650.
Moore, K. W., Sher, B. T., Sun, Y. H., Eakle, K. A. & Hood, L. 1982 *Science, Wash.* **215**, 679–682.
Morré, D. J., Schirrmacher, V., Robinson, P., Hess, K. & Franke, W. W. 1979 *Expl Cell Res.* **119**, 265–275.
Nathenson, S. G. & Cullen, S. E. 1974 *Biochim. biophys. Acta* **344**, 1–25.
Nathenson, S. G., Uehara, H., Ewenstein, B. M., Kindt, T. J. & Coligan, J. E. 1981 *A. Rev. Biochem.* **50**, 1025–1051.
Örn, A., Goodenow, R. S., Hood, L., Brayton, P. R., Woodward, J. G.. Harmon, R. C. & Frelinger, J. A. 1982 *Nature, Lond.* **297**, 415–417.
Orr, H. T., Lopez de Castro, J. A., Parham, P., Ploegh, H. L. & Strominger, J. L. 1979 *Proc. natn. Acad. Sci. U.S.A.* **76**, 4395–4399.
Owen, M. J., Kissonerghis, A. M. & Lodish, H. L. 1980 *J. biol. Chem.* **255**, 9678–9684.
Parham, P., Alpert, B. N., Orr, H. T. & Strominger, J. L. 1977 *J. biol. Chem.* **252**, 7555–7567.
Pearse, B. M. F. & Bretscher, M. S. 1981 *A. Rev. Biochem.* **50**, 85–101.
Peterson, P. A., Rask, L. & Lindblom, J. B. 1974 *Proc. natn. Acad. Sci. U.S.A.* **71**, 35–39.
Ploegh, H. L., Cannon, L. E. & Strominger, J. L. 1979 *Proc. natn. Acad. Sci. U.S.A.* **76**, 2273–2277.
Ploegh, H. L., Orr, H. T. & Strominger, J. L. 1981 *a Cell* **24**, 287–299.
Ploegh, H. L., Orr, H. T. & Strominger, J. L. 1981 *b J. Immunol.* **126**, 270–275.
Pober, J. S., Guild, B. S. & Strominger, J. L. 1978 *Proc. natn. Acad. Sci. U.S.A.* **75**, 6002–6006.
Pober, J. S., Guild, B. C., Strominger, J. L. & Veatch, W. R. 1981 *Biochemistry, Wash.* **20**, 5625–5633.
Pober, J. S. & Strominger, J. L. 1981 *Nature, Lond.* **289**, 819–821.
Rask, L., Lindblom, J. B. & Peterson, P. A. 1974 *Nature, Lond.* **249**, 833–834.
Robb, R. J., Terhorst, C. & Strominger, J. L. 1978 *J. biol. Chem.* **253**, 5319–5324.
Rothbard, J. B., Hopp, T. P., Edelman, G. M. & Cunningham, B. A. 1980 *Proc. natn. Acad. Sci. U.S.A.* **77**, 4239–4243.
Rothman, J. E. & Lodish, H. F. 1977 *Nature, Lond.* **269**, 775–780.
Sege, K., Rask, L. & Peterson, P. A. 1981 *Biochemistry, Wash.* **20**, 4523–4530.
Sood, A. K., Pereira, D. & Weissman, S. M. 1981 *Proc. natn. Acad. Sci. U.S.A.* **78**, 616–620.
Steinmetz, M., Frelinger, J. G., Fisher, D., Hunkpiller, T., Pereira, D., Weissman, S. M., Uehara, H., Nathenson, S. & Hood, L. 1981 *Cell* **24**, 125–134.
Steinmetz, M., Winoto, A., Minard, K. & Hood, L. 1982 *Cell* **28**, 489–498.
Suggs, S. V., Wallace, R. B., Hirose, T., Kawashima, E. H. & Itakura, K. 1981 *Proc. natn. Acad. Sci. U.S.A.* **78**, 6613–6617.
Tabas, I. & Kornfeld, S. 1978 *J. biol. Chem.* **253**, 779–786.
Tabas, I., Schlessinger, S. & Kornfeld, S. 1978 *J. biol. Chem.* **253**, 716–722.
Tarentino, A. L. & Maley, F. 1974 *J. biol. Chem.* **249**, 811–817.
Tartakoff, A., Hoessli, D. & Vassalli, P. 1981 *J. molec. Biol.* **150**, 525–535.
Terhorst, C., Robb, R., Jones, C. & Strominger, J. L. 1977 *Proc. natn. Acad. Sci. U.S.A.* **74**, 4002–4006.
Trägårdh, L., Rask, L., Wiman, K., Fohlman, J. & Peterson, P. A. 1980 *Proc. natn. Acad. Sci. U.S.A.* **77**, 1129–1133.
Vitetta, E. S. & Uhr, J. W. 1975 *J. Immunol.* **115**, 374–381.
Walsh, F. S. & Crumpton, M. J. 1977 *Nature, Lond.* **269**, 307–311.
Walter, P. & Blobel, G. 1981 *J. Cell Biol.* **91**, 557–561.
Warren, G. 1980 In *Comprehensive biochemistry* (ed. R. Michel & B. Finean), vol. 1, pp. 215–257. Amsterdam: Elsevier/North-Holland.
Zinkernagel, R. M. & Doherty, P. C. 1974 *Nature, Lond.* **148**, 701–702.

Phil. Trans. R. Soc. Lond. B **300**, 173–184 (1982)

Printed in Great Britain

The role of subcompartments of the Golgi complex in protein intracellular transport

By A. M. Tartakoff

Department of Pathology, University of Geneva School of Medicine, Centre Médical Universitaire,
1, rue Michel-Servet, CH-1211 Geneva 4, Switzerland

[Plates 1 and 2]

The functioning of the Golgi complex in protein intracellular transport is most simply understood in terms of its being composed of a sequence of functionally distinct subcompartments. For example, the influence of perturbation of cellular Na^+–K^+ balance on the transport of secretory and membrane glycoproteins is to greatly slow their passage from relatively proximal to relatively distal subcompartments. To further the understanding of the nature of these subcompartments a rat IgM myeloma has been subjected to analytical subcellular fractionation. Fractions selectively enriched in distinct Golgi-associated activities have been prepared and their membrane proteins compared with those of rough microsomal fractions. The subfractionation is extensive and suggests the possibility of obtaining well resolved Golgi subfractions. Myeloma cells stained intracellularly with Concanavalin A– and wheatgerm agglutinin–peroxidase conjugates show distinct labelling patterns. Concanavalin A stains the entirety of the rough endoplasmic reticulum as well as the proximal face of the Golgi stack. Wheatgerm agglutinin stains the distal face of the stack of Golgi cisternae. The staining patterns are not due to immunoglobulin as they are also observed in myeloma variants that fail to synthesize immunoglobulin.

Introduction

(a) Background

The transport of secretory proteins from the rough endoplasmic reticulum (r.e.r.) to the cell surface is mediated by smooth-surfaced vesicles and the elongated stacked cisterna of the Golgi complex. These several membranes interact successively in such a way as to accomplish rapid and efficient transport, allowing a given protein to reach the surface roughly 7–20 min after biosynthesis. The mechanisms operating along this pathway are largely obscure and in fact we even lack systematic compositional information about these distinct or partly distinct classes of membranes (Tartakoff 1980, 1982).

The importance of analysing these Golgi-associated transport operations is heightened by the realization that transport is highly selective: in addition to the secretory proteins, lysosomal proteins and numerous membrane proteins also pass by way of the Golgi complex. Many of these proteins deviate from the secretory path and avoid delivery to the cell surface.

It has been clear both to anatomists and histochemists that there is considerable heterogeneity among the several cisternae of the Golgi complex. For example, more proximal cisternae (oriented toward the transitional elements of the r.e.r.) can be selectively stained by over-osmication, while relatively distal cisternae are histochemically positive for uridine disphosphatase. It is my conviction that to understand the mechanism of transport we must learn the composition of the membranes of the Golgi complex and its subcompartments, we must learn to separate the subcompartments and ultimately to put them back together again.

[65]

(b) Choice of tissue

It is customary to isolate subcellular fractions from the rat liver, yet especially for the particular job at hand the liver is not altogether suitable. The cell population of the liver is mixed, its Golgi complex is not highly organized, and large quantities of a potential low-density contaminant – the smooth endoplasmic reticulum – are present. Moreover, the hepatocyte has the unfortunate habit of producing a wide range of secretory proteins. The presence of such abundant secretory protein content within subcellular fractions poses an overwhelming problem: although procedures are known for eliminating such content, even the staunchest advocates of such procedures must admit that the extractions are incomplete (Howell & Palade 1982; Kreibich & Sabatini 1974). Thus in an isolated fraction it is impossible to discriminate rigorously between those proteins that are part of the cell's transport machinery and those that are being transported. Moreover, the secretory lipoproteins diminish the inherent density of the elements they traverse. In fact, this latter property of the liver has been exploited so as to recover highly purified Golgi subfractions of very low buoyant density (Morré 1971; Ehrenreich et al. 1973), which have been of major importance, especially for the study of the terminal sugar transferases. Nevertheless it is my feeling that we have to get away from such a preparative approach to the subcellular fractionation of the Golgi complex and that an analytical, all-inclusive approach would ultimately prove to be more powerful (de Duve 1971).

For these several reasons I have undertaken to study the Golgi complex and Golgi-enriched subcellular fractions of a myeloma. Homogeneous cell populations can be obtained in gram quantities and are ideally suited for rapid in vitro manipulations, pulse-chase labelling, viral infection, etc. Only a single secretory protein is present, thereby greatly reducing the danger of the secretory content's obscuring the analysis of the membrane proteins of subcellular fractions. Furthermore, the cells are so-called 'non-regulated' secretory cells (i.e. they secrete continuously without storing their secretory product), and are therefore relatively simple from the cytological point of view. In addition, since the biosynthesis of the carbohydrate of immunoglobulin (Ig) has been extensively studied, one can use selected stages of carbohydrate maturation as potential markers of compartments involved in intracellular transport. We work with an IgM-secreting rat myeloma (IR202, obtained from H. Bazin) because IgM is rather rich in carbohydrate and because by working with a rat tissue we expect to be able to take advantage of much of what is known of the enzymology and antigenicity of other rat tissues, including the liver. Most of the experiments I shall describe in this paper make use of myeloma cells grown as an ascitic tumour. In this form they devote about 12 % of their protein biosynthesis (10 min, [^{35}S]methionine pulse label) to secretory IgM production. We have no evidence for the presence of surface Ig on these cells. The cells do not contain visible viral particles.

(c) Specific goals

I wish to exemplify several distinct experimental approaches to the analysis of the structure and function of the Golgi complex of this myeloma. The first approach is pharmacological and makes use of the carboxylic ionophore monensin to perturb the Golgi complex. The second involves performing analytical subcellular fractionation in such a way as to spread Golgi elements in density gradients, and then examining the polypeptide composition of such subcellular fractions. The Golgi-enriched fractions are compared with rough microsomal fractions and with comparable fractions isolated from the rat liver. I shall then turn to the use of ultrastructural cytochemistry to localize lectin-binding sites, which are characteristic of the r.e.r. or Golgi subcompartments. Throughout, the emphasis will be on the divisibility of the Golgi complex into subcompartments and their possible role(s) in intracellular transport.

Pharmacological approach

The idea of attempting to perturb Golgi structure and function by manipulating cellular Na^+ and K^+ levels grew from some incomplete and scattered observations in the ultrastructural literature and from the realization that many cells must be able to tolerate alterations in Na^+–K^+ balance. The most dramatic agent employed is the carboxylic ionophore monensin, which appears to be universally effective on eukaryotic cells. I shall summarize its principal effects, which are published (Tartakoff 1983), and illustrate a few in the system at hand.

In the presence of 0.1–1.0 μM monensin, cells rapidly lose K^+ and gain a roughly stoichiometric amount of Na^+. Intracellular pH changes are brief and modest, there is no effect on ATP levels or Ca^{2+} efflux, and protein synthesis continues at a normal or near-normal rate. What is especially striking is that within no more than a few minutes the intracellular transport of both secretory and plasma membrane proteins ceases. Such an effect on IgM secretion can easily be observed when a 5 min [^{35}S]methionine pulse is followed by a 2 h chase incubation in control or monensin-containing medium. The presence of monensin reduces the amount of extracellular immunoprecipitable [^{35}S]Ig by *ca.* 90 %. Both autoradiographic and subcellular fractionation experiments show that in the presence of monensin newly synthesized proteins continue to exit from the r.e.r. at a normal rate and overaccumulate within the Golgi complex.

Available information indicates that each Ig heavy chain bears five or six asparagine-linked carbohydrate units and that the oligosaccharides of secreted Ig are fully resistant to endoglucosaminidase H. This resistance is normally acquired during passage through the Golgi complex due to the concerted action of glycosidases and terminal sugar transferases. It is worth pausing to recognize the high efficiency of this oligosaccharide maturation process, because an individual pentameric IgM molecule in the r.e.r. bears a total of 50–60 immature (endoglucosaminidase H-sensitive) oligosaccharide units.

We have enquired whether the IgM that is caused to accumulate in the presence of monensin has completed the normal maturation of its carbohydrate. The protocol that we have employed is a [^{35}S]methionine pulse–chase experiment, where the chase is either in control medium or in medium supplemented with monensin, or with the phenylhydrazone uncoupler of oxidative prosphorylation, CCCP, which arrests exit of Ig from the r.e.r. For comparison, samples of cells labelled in the presence of tunicamycin are also analysed. Each sample of immunoprecipated Ig is reduced and alkylated, divided in two, and incubated without or with endoglucosaminidase H. Figure 1 illustrates the sodium dodecyl sulphate polyacrylamide gel electrophoretic mobility of the labelled heavy chains, as revealed by autoradiography. In the pulse-labelled sample the mobility is slightly in advance of that of secreted heavy chains. All the chase intracellular samples are of similar mobility to the pulse-labelled sample (except after tunicamycin treatment, used to indicate the mobility of carbohydrate-free heavy chains). Upon endoglocosaminidase treatment, *all* the intracellular samples can be seen to be sensitive to their enzyme, as indicated by their mobility becoming almost as rapid as that of carbohydrate-free chains. The extracellular sample is unaffected. Since the control chase sample is fully sensitive, the intracellular pool of Ig with mature carbohydrate must be quite small.

These observations, and others, show that secretory Ig carbohydrate maturation is greatly retarded in the presence of the ionophore. We have obtained analogous data bearing on the transport and oligosaccharide maturation of two plasma-membrane proteins, surface Ig and histocompatibility antigens. Because a number of control experiments show that the terminal sugar transferases are still active, we consider this effect to be ascribable to topographic, not

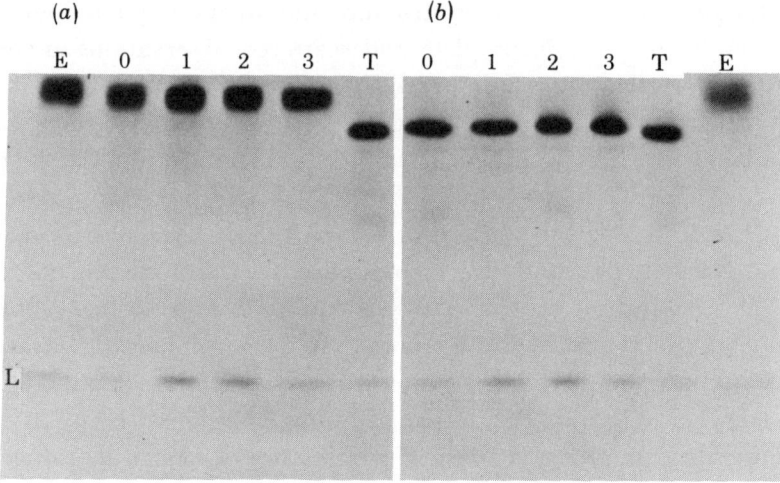

FIGURE 1. Evaluation of secretory Ig heavy chain carbohydrate maturation by use of endoglucosaminidase H. Ascitic IR 202 myeloma cells were pulse-labelled for 5 min with [^{35}S]methionine, washed, and either chilled (0) or returned to non-radioactive medium for 2 h at 37 °C. Among the reincubated samples, one was incubated without the addition of any perturbant of intracellular transport (1), one was incubated in the presence of 1 μM monensin (2), and one was incubated in the presence of 10 μM CCCP (3). An additional sample of cells (T) was preincubated for 1 h in tunicamycin (1 μg ml^{-1}), pulse-labelled, washed, and chilled at once. In each case, intracellular [^{35}S]Ig was recovered by immunoprecipitation and in condition (1) an extracellular sample (E) was recovered as well. Each sample was divided into halves, reduced and alkylated and incubated overnight at pH 5.5, 37 °C, without (a) or with (b) endoglucosaminidase H (30 μM ml^{-1}). Samples were then analysed by standard sodium sulphate polyacrylamide gel electrophoresis – autoradiography procedures. Because the enzyme removes high-mannose carbohydrate units, but not mature complex units, it allows a rough discrimination between polypeptides bearing the two classes of carbohydrate unit. It is clear that all the intracellular samples are sensitive to the enzyme, which implies that they bear immature carbohydrate units. By contrast, the extracellular sample is resistant. After enzyme treatment the intracellular samples still have a slightly slower mobility than the carbohydrate-free sample (T), presumably because the enzyme leaves the most internal glucosamine residue on the polypeptide chain and the heavy chains of IgM bear a total of five or six carbohydrate units. L, Ig light chains.

enzymological, considerations. Thus we are forced to think in terms of the Golgi's being composed of at least a pair of subcompartments: a proximal one that newly synthesized proteins can continue to enter in the presence of monensin, and a more distal one that is relatively inaccessible in the presence of monensin and in which maturation to endoglucosaminidase H resistance normally occurs. Because a number of authors have employed monensin to study the intracellular transport of other macromolecules it is possible to ascribe additional post-translational modifications to these two Golgi subcompartments, defined by reference to monensin's action. Thus a 'confined function model' of the Golgi complex can be considered in which the proximal compartment is the site of phosphorylation of lysosomal enzymes and lipid addition to viral envelope glycoproteins, whereas the distal compartment is responsible for the completion of the oligosaccharide maturation of asparagine-linked oligosaccharides and the sulphation of proteoglycans (Tartakoff 1982, 1983).

In addition to interrupting intracellular transport, monensin has another striking effect: it causes a massive dilation of Golgi elements, which nevertheless remain closely adherent one to the next. It is not clear what causal relation, if any, exists between this dilation and the interruption of transport; however, the mere ultrastructural observation in its own right implies that the membranes of Golgi cisternae have a unique property (probably related to ion permeability) that distinguishes them from the membranes of the endoplasmic reticulum. This unique property deserves to be identified.

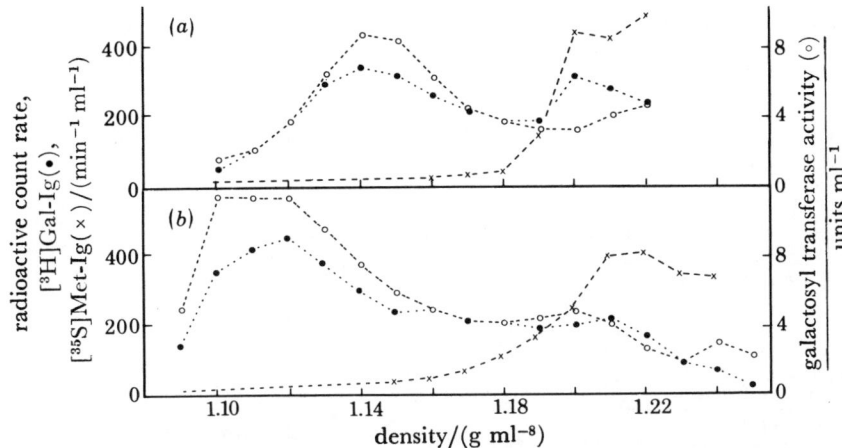

FIGURE 2. Comparative subcellular fractionation of myeloma cell postnuclear supernatants on (a) mannose-containing and (b) sucrose-containing gradients. Cells were labelled for 1 min with [^{35}S]methionine or for 5 min with [^{3}H]galactose and then mixed, homogenized, and briefly centrifuged to remove debris. The resulting supernatant was fractionated in parallel on two gradients, centrifuged overnight to equilibrium. Radioactive Ig was immunoprecipitated from each fraction and galactosyl transferase activity was measured as in Howell *et al.* (1978). In both gradients the [^{35}S]Ig is entirely recovered at densities greater than 1.18 g ml^{-1}. Coincident peaks of [^{3}H]Ig and galactosyl transferase are observed in lower-density regions. It is not known why a proportion of these activities is also recovered in denser fractions; it is possibly because the Golgi elements within which they reside bear extensive, clathrin-containing coats (coated vesicles normally sediment to *ca.* 55% sucrose). The higher equilibrium density of the Golgi markers in mannose gradients is suggested to be a reflection of the ability of mannose (unlike sucrose) to penetrate the membranes.

SUBCELLULAR FRACTIONATION

In light of the well established structural and histochemical heterogeneity of the stack of Golgi cisternae, and with the monensin data in mind, we have sought to isolate Golgi-enriched subcellular fractions and if possible subfractions. Our first concern has been to ensure that bound ribosomes remain attached to the r.e.r. and to this end we have taken two precautions: (1) just before chilling and homogenization, cells are treated with cycloheximide, which is also included in the homogenate, and (2) the fractionation is performed in the presence of Mg^{2+}. After the elimination of nuclei and debris, as a basic assessment of the adequacy of the conditions of isopycnic sedimentation we have sought to localize three parameters: (1) Ig labelled with a 1 min pulse of [^{35}S]methionine (to mark the rough microsomes), (2) Ig labelled with a 5 min pulse of [^{3}H]galactose (which should mark Golgi elements), and (3) galactosyl transferase activity measured with ovomucoid as acceptor (again a Golgi marker). We have performed such fractionation on sucrose gradients, yet the procedure of choice for the time being involves fractionation on mannose gradients. There are several motivating considerations: (1) in sucrose gradients the peaks of galactosyl transferase are routinely quite broad, (2) the published electron micrographs of Golgi fractions isolated in the presence of sucrose suggest that they are osmotically compressed (Ehrenreich *et al.* 1973), (3) if one performs subcellular fractionation in the presence of non-penetrating solutes, one must expect that both the size and the intrinsic density of the particles will influence their isopycnic sedimentation (de Duve 1965), and (4) it is known

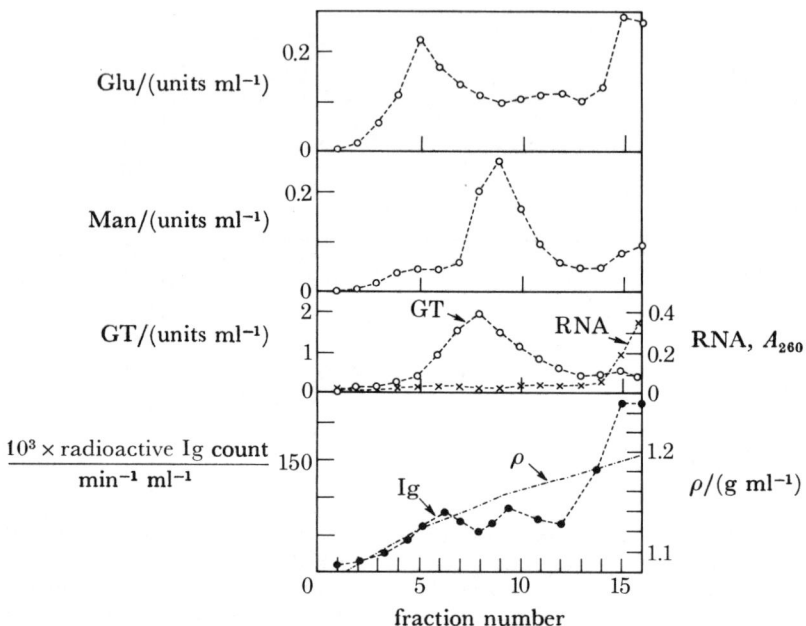

FIGURE 3. Fractionation of total microsomal fractions on mannose gradients. Non-radioactive ₍ ₎itic myeloma cells or cells labelled for 60 min with [^{35}S]methionine were homogenized and centrifuged to eliminate nuclei, mitochondria (detected by succinate dehydrogenase activity), and lysosomes (detected by endocytosed peroxidase activity). The resulting 'post-mitochondrial supernatants' were sedimented to collect total microsomal fractions, which were fractionated by overnight centrifugation on parallel continuous mannose gradients. The gradients were analysed for immunoprecipitable [^{35}S]Ig, RNA, galactosyl transferase activity (GT), mannosidase activity (Man) and glucosidase activity (Glu). Unlike pulse-labelled [^{35}S]Ig (figure 2), the 60 min labelled [^{35}S]Ig has gained access to compartments of relatively low density and accumulates sufficiently to give a pair of peaks centred near 1.13 and 1.16 g ml^{-1}. Three distinct Golgi-associated enzyme activities are also recovered in the upper part of the gradient. The non-coincidence of peaks of GT activity and [^{35}S]Ig indicates that a partial resolution of Golgi elements has in fact been achieved.

that mannosidase activity is present within Golgi membranes (Tabas & Kornfeld 1978, 1979 Tulsiani *et al.* 1977), and thus the mannosidase-bearing or more distal Golgi elements must be permeable to mannose; if not, they would swell. Figure 2 illustrates comparative isopycnic sedimentation on mannose and sucrose gradients. It is clear that in both cases the rough microsomes can be effectively eliminated, that the resolution of Golgi-associated activities is superior in the mannose gradient, and that their density is greater than in the sucrose gradient, presumably owing to the permeation of mannose. As anticipated, the peak of transferase activity is coincident with Ig pulse-labelled with [^{3}H]galactose. There is no indication of rupture and leakage of labelled Ig. Moreover, electron microscopic examination of the fractions centered at 1.13 or 1.16 g ml^{-1} shows that the fractions are largely composed of closed smooth-surfaced vesicles. There is no indication that they have undergone osmotic compression.

We have made use of such mannose gradients to analyse total microsomal fractions from cells labelled for 60 min with [^{35}S]methionine or [^{3}H]2-D-mannose. When labelled Ig is recovered by immunoprecipitation, three peaks are seen, centred at densities of *ca.* 1.13, *ca.* 1.16, and greater than 1.19 g ml^{-1} (figure 3). Essentially all of the labelled Ig recovered from these gradient regions is sensitive to endoglucosaminidase H. Hence they correspond to relatively

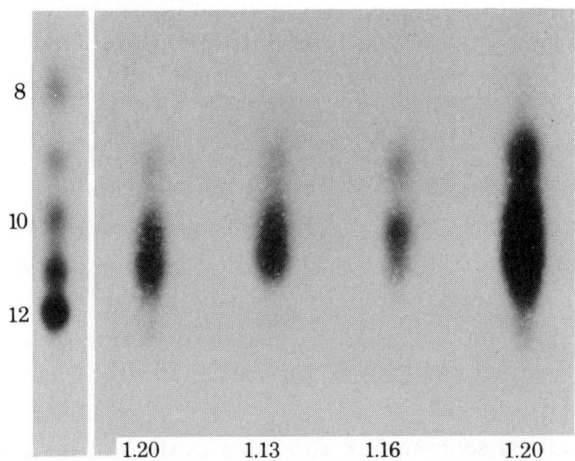

FIGURE 4. Analysis of [^{3}H]mannose-labelled Ig oligosaccharides recovered from mannose gradients. Myeloma cells were labelled for 60 min with [^{3}H]2-D-mannose in the presence of glucose (0.1 mg ml^{-1}) and fractionated to provide three peaks of labelled Ig cosedimenting with the peaks of [^{35}S]Ig illustrated in the bottom panel of figure 3. Each sample of [^{3}H]Ig recovered by immunoprecipitation was reduced, alkylated, treated with endoglucosaminidase H, desalted and fractionated by thin-layer chromatography on silica gel followed by fluorography. At the left is a family of [^{3}H]mannose-labelled oligosaccharides released from dolichol by weak acid hydrolysis and endoglucosaminidase H. The sizes of the species were assessed by comparison with dolichol and viral glycoprotein-derived standards obtained from Dr M. Snider (Massachusetts Institute of Technology) and Dr R. Datema (Giessen). A labelled [^{3}H]oligosaccharide sample recovered from the bottom of the gradient (1.20 g ml^{-1}) is run twice for comparison with material recovered from fractions centred at 1.13 and 1.16 g ml^{-1}. These rough microsomal oligosaccharides range in size from 11 to 9 sugar residues and are therefore considered to include GlcNAc$_1$Man$_9$Glc$_1$, GlcNAc$_1$Man$_9$ and GlcNAc$_1$Man$_8$. As suggested by Godelaine *et al.* (1982), species such as Glc$_1$Man$_8$GlcNAc$_1$ may also be present. The principal oligosaccharide obtained from [^{3}H]Ig of density 1.13 and 1.16 g ml^{-1} is somewhat smaller than the most abundant rough microsomal species and is therefore thought to lack glucose residues. In addition, a faint spot in the nonasaccharide region continues to be present.

proximal stations in Ig intracellular transport. We have enquired to what extent these three samples of labelled Ig can be ordered along the path of intracellular transport. To analyse the size of incomplete Ig oligosaccharides recovered from each gradient region, we have reduced and alkylated [^{3}H]mannose-labelled immunoprecipitates, treated with endoglucosaminidase H, and fractionated the released oligosaccharides by thin-layer chromatography (Holmes & O'Brien 1979). Figure 4 shows a fluorogram of such a chromatogram. At the level of rough microsomes, the sensitive oligosaccharides range in size from 11 to 9 sugar residues, which we provisionally consider to be GlcNAc$_1$Man$_9$Glc$_1$, GlcNAc$_1$Man$_9$, and GlcNAc$_1$Man$_8$. The last of these species has also been detected in thyroglobulin recovered from thyroid rough microsomes (Godelaine *et al.* 1982), thus showing that one mannose residue (as well as the glucose residues) can be removed before entry into the Golgi complex. In the two lighter density gradient regions, the largest of these species is less evident, suggesting that glucose removal has been completed. Nevertheless, the principal conclusion to be drawn is that extensively trimmed species (e.g. $GlcNAc_1Man_5$ or $GlcNAc_1Man_3$) must be of such transient existence that they cannot serve as markers of subcompartments, given the currently available fractionation procedures. This point might be re-explored making use of a myeloma variant that lacks selected terminal sugar transferases.

Figure 3 also shows the distribution of RNA, which is highly concentrated at the bottom of the gradient, and two other Golgi-associated enzyme activities implicated in the processing of oligosaccharides of glycoproteins: glucosidase (which is also recovered in the rough microsomal fraction) and mannosidase, both measured at pH 6 with p-nitrophenyl-α-D-glycoside substrates. We expect that these assay conditions detect principally the glucosidase AB of Grinna & Robbins (1979) and the α-1-3, 1-6 mannosidase of Tabas & Kornfeld (1979). Both activities have previously been detected in preparative Golgi fractions of rat liver, and the glucosidase has been detected at high specific activity in rat liver rough microsomes.

We conclude that upon exit from the r.e.r., Ig gains access to lower density glycosidase-rich gradient regions ranging in density from 1.12 to 1.17 g ml^{-1}, acquires galactose and possibly other terminal sugars within a compartment of density 1.15 g ml^{-1}, and quickly exits from the cell. We have accomplished a more extensive partial resolution of Golgi elements than has previously been reported; however, still more powerful analytical fractionation tools are clearly needed.

The subcellular fractionation scheme has also been evaluated by sodium dodecyl sulphate (SDS) polyacrylamide gel electrophoresis in the hope of seeing whether, despite the persistent overlap of Golgi-associated activities, certain generalizations might be made with respect to the polypeptide composition of these membranes. In order to simplify these gel patterns, somewhat, we have also treated individual gradient fractions with pH 11.2, 0.1 M Na_2CO_3 (Howell & Palade 1982) to extract a maximum of extrinsic protein, ribosomes, and content before loading the gels. Silver nitrate-stained gels are shown as figure 5. What is clear, concentrating on the pH 11 residue, is that the rough microsomes (best represented by the last two lanes on the right) are distinctly different from the continuum of fractions that range through the 1.16 and 1.13 g ml^{-1} density regions toward the top of the gradient. Moreover, there are certain bands that span most of the fractions that we have reason to identify as including Golgi elements (e.g. the band with molecular mass greater than 200 kDa indicated by an asterisk), and within this broad 'Golgi continuum' a number of subregions can be identified by their characteristic polypeptides. Although the resolution of such one-dimensional gels is limited, certain bands appear to be common to both the Golgi-enriched regions and the rough microsomal fraction.

Are these Golgi-enriched gradient regions representative of the Golgi complex of another cell? Which are the polypeptides essential to the basic transport operations of the Golgi complex, as opposed to the particular operations of a single cell type? To attempt to answer these questions, three myeloma gradient fractions (density 1.13, 1.16 and over 1.19 g ml^{-1}) have been extracted at pH 11 and compared with liver fractions extracted in parallel. The liver rough microsomal fraction was isolated by a procedure designed to maximize yield without compromising purity (Adelman *et al.* 1973); however, the Golgi-enriched fractions are all preparative, selected according to their very low density (due especially to their content) and the high specific activity of terminal sugar transferases (Ehrenreich *et al.* 1973; Leelevathi as cited by Tabas & Kornfeld 1979). As indicated by the asterisks in figure 6, a number of bands are shared between tissues. These obviously are the best candidates for further investigation and will be of particular interest with respect to raising antisera against the underlying components of the organelle.

LECTIN-BINDING PROTEINS

I wish now to shift emphasis and instead of looking at the totality of the proteins associated with the Golgi complex I shall focus on two largely distinct classes of proteins, those that bind Concanavalin A (Con A) (and therefore contain α-linked mannose-rich oligosaccharides unsubstituted at C-3, C-4 and C-6 (Kornfeld & Ferris 1975)) and those that bind wheatgerm

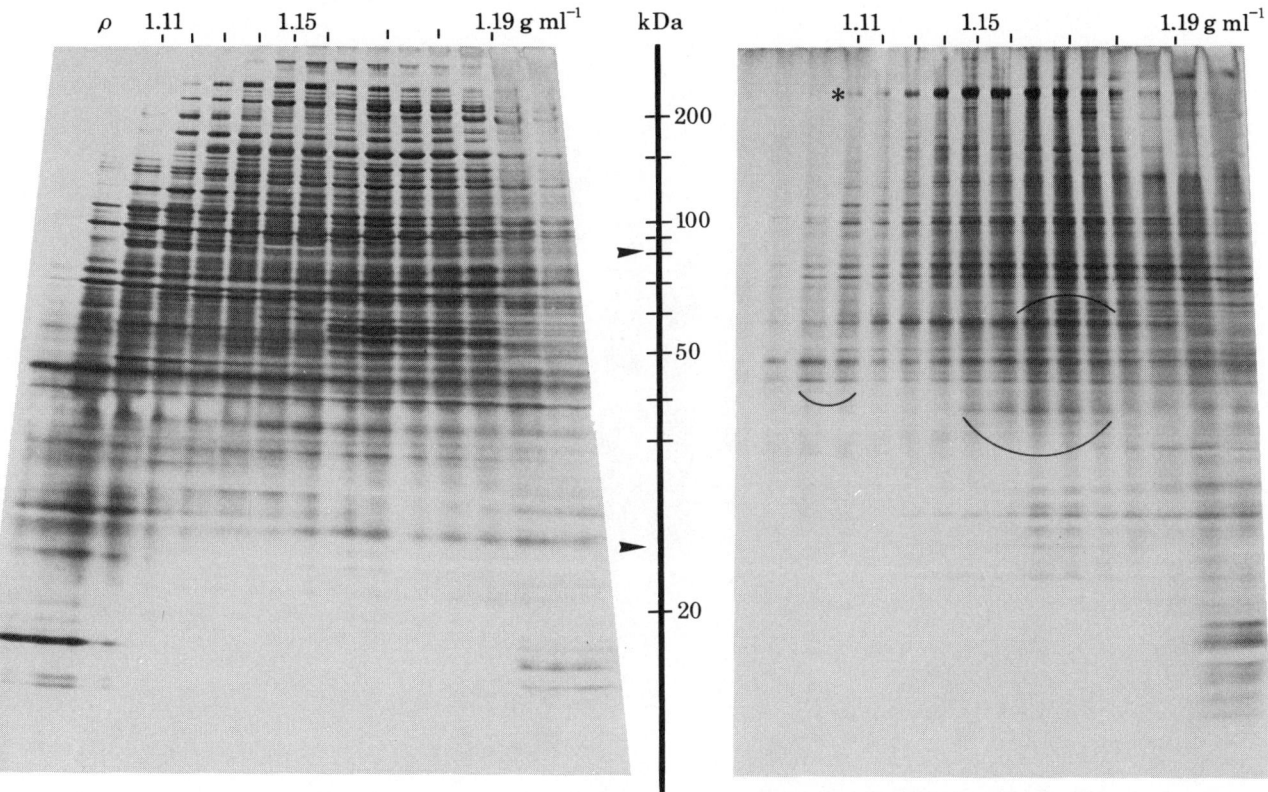

FIGURE 5. SDS polyacrylamide gel electrophoretic analysis of total microsomal fractions after resolution on continuous mannose gradients. Gradients identical to those illustrated in figure 3 were sampled at regular intervals and the successive fractions were analysed by gel electrophoresis followed by staining with silver nitrate. In the panel at the right the samples were extracted at pH 11.2 according to Howell & Palade (1982) before analysis. The mobilities of Ig heavy and light chains are indicated by the arrowheads along the molecular mass axis. These Ig chains are not conspicuous on a mass basis, although after biosynthetic labelling they predominate. In the right-hand panel the asterisk indicates a major protein common to all gradient regions including Golgi elements. The arcs indicate regions of the gradients with characteristic polypeptide composition, which are therefore thought to correspond to distinct membrane populations.

agglutinin (WGA) (and therefore bear terminal sialic acid or N-acetylglucosamine (GlcNAc) residues (Bhavanandan & Katlic 1979)). It has recently been reported that such lectins conjugated to fluorochromes give distinct staining patterns of the cytoplasm of fibroblasts and that WGA is selective for the Golgi complex, as judged both from the localization of fluorescence and the influence of monensin on the staining pattern (Virtanen et al. 1980).

We have repeated these observations on the rat myeloma and extended them to the ultrastructural level, making use of the saponin-permeabilization procedure originated by Ohtsuki *et al.* (1978). When cytocentrifuged, air-dried, fixed cells are stained with fluorescein or rhodamine-conjugated lectins, one can readily show that essentially all of the cell cytoplasm is Con A-positive whereas the staining with WGA is largely restricted to a single perinuclear zone coincident with histochemically detected uridine diphosphatase.

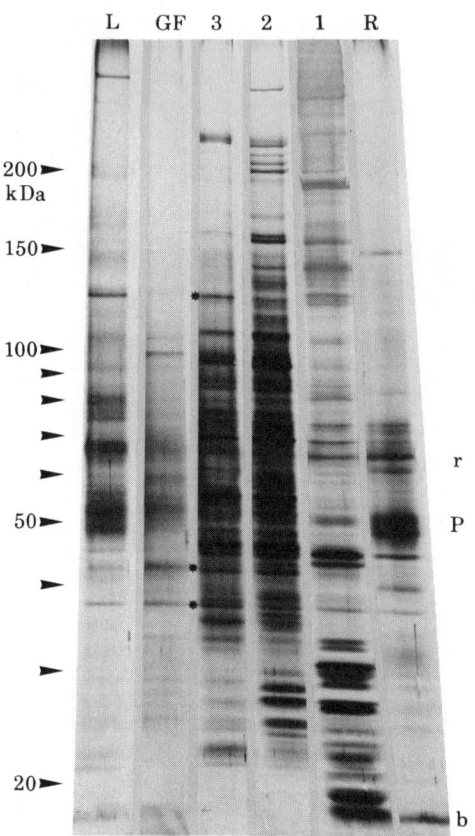

FIGURE 6. SDS polyacrylamide gel electrophoretic comparison of the protein composition of microsomal sub-fractions of the myeloma with corresponding conventional rat liver fractions, all after extraction at pH 11.2 Samples (1), (2), and (3) are derived from the mannose gradient illustrated in figure 3, (1) from the bottom of the gradient, (2) from the region at 1.16 g ml^{-1}, and (3) at 1.13 g ml^{-1}. The sample labelled R is a liver rough microsomal fraction, GF a Golgi fraction isolated (without ethanol intoxication) according to Ehren-reich *et al.* (1973) and L a Golgi fraction isolated according to Leelavathi as reported by Tabas & Kornfeld (1979). At the extreme right are indicated the expected mobilities of several rat liver microsomal proteins, the ribophorins (r), the cytochromes P$_{450}$ (P) and cytochrome b$_5$ (b). The stars indicate proteins that are largely shared among the myeloma and liver 'Golgi fractions'.

At the ultrastructural level the staining patterns are more striking. We use lectin conjugates coupled through the carbohydrate moiety of peroxidase (Nakane & Kawaoi 1974) and find that permeabilization is very extensive because all cells (and, in the case of Con A–horseradish peroxidase (HRP), all r.e.r. profiles) are labelled. In figure 7, plate 1, one sees a typical cell stained with Con A–HRP. The reaction product is restricted to the r.e.r. and the perinuclear cisterna, where one may note the presence of (unstained) nuclear pores. The extensive staining of the r.e.r. is consistent with staining of myeloma homogenates reported by Hirano *et al.* (1972)

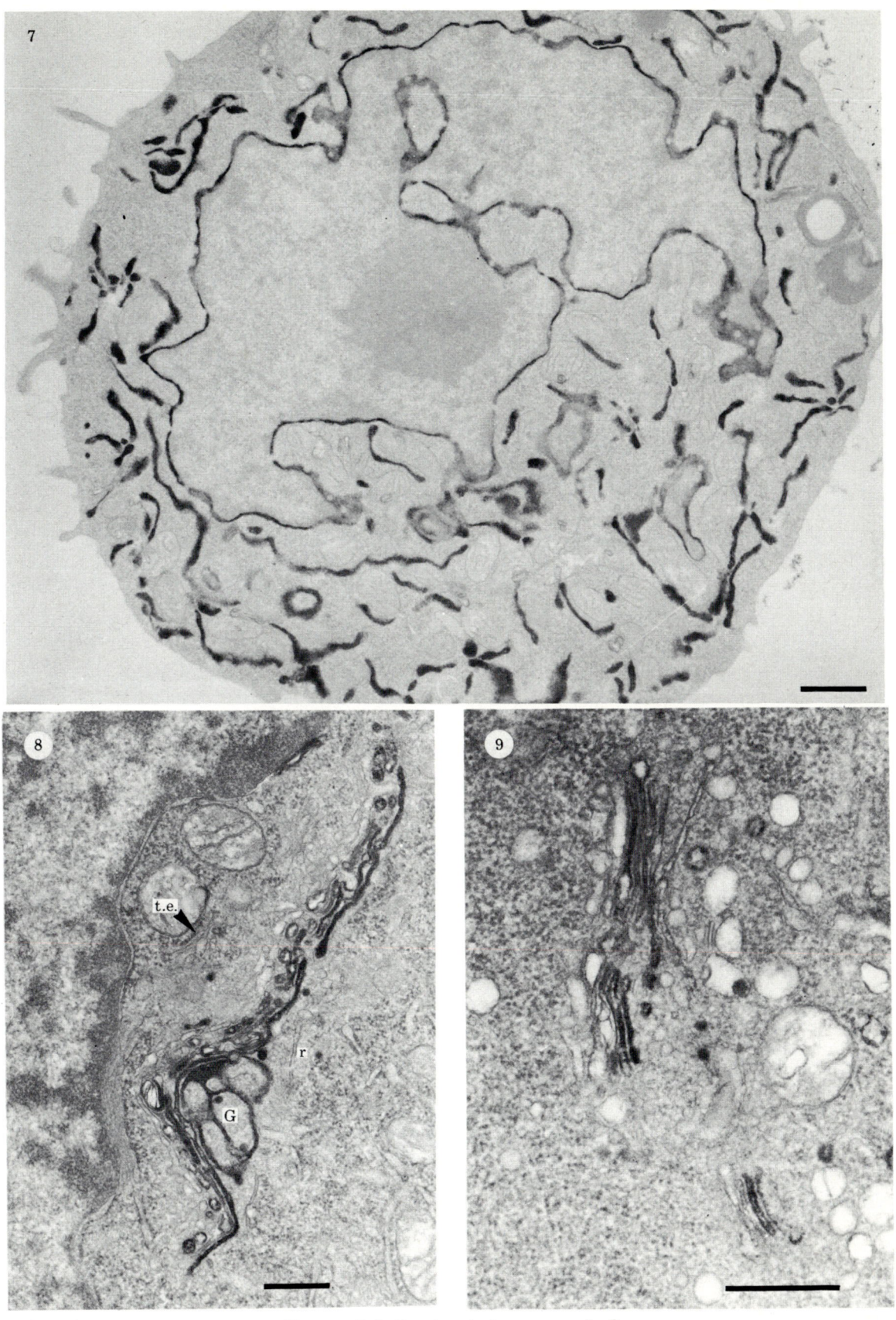

FIGURES 7–9. For description see overleaf.

Phil. Trans. R. Soc. Lond. B, volume **300**

Tartakoff, plate 2

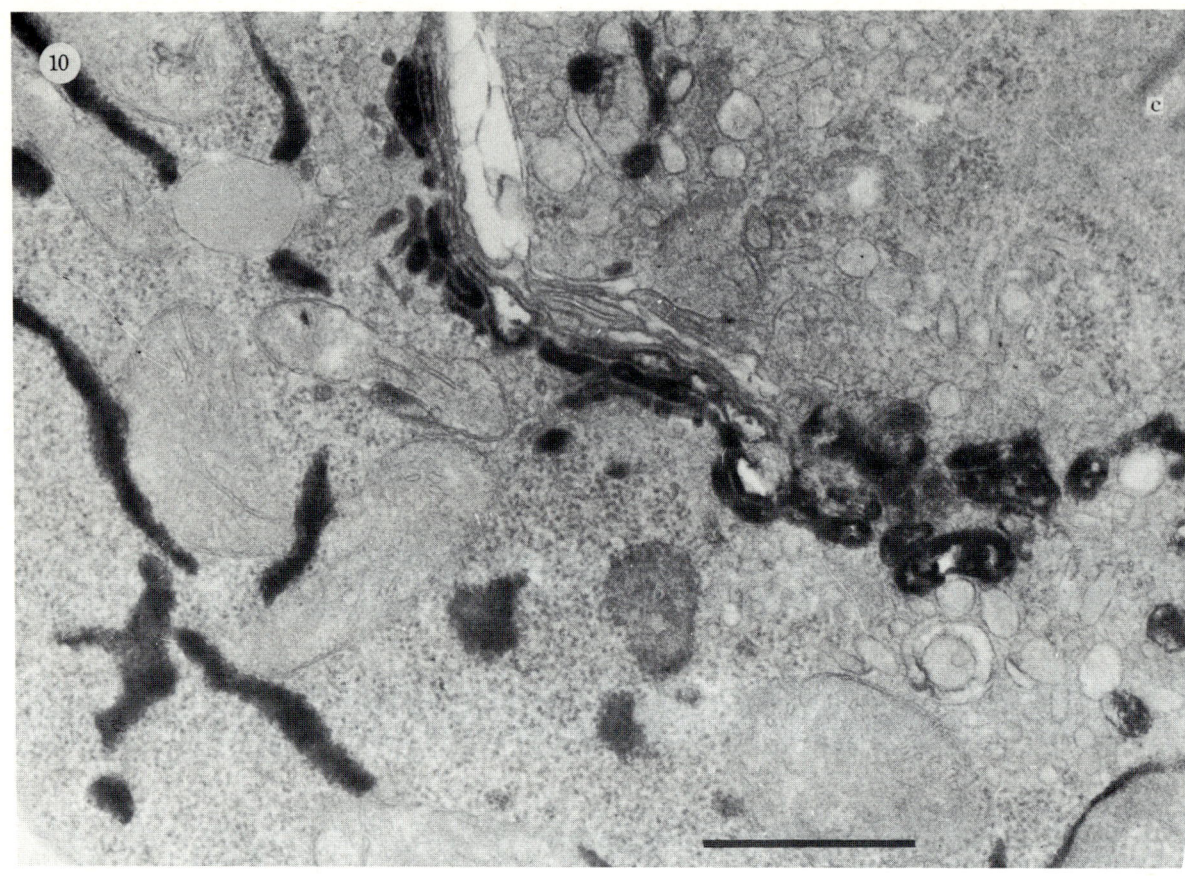

FIGURES 7–10. Thin sections of the rat myeloma (figures 7, 8 and 10) or the non-producing mouse myeloma Sp 2/0 Ag 14 (figure 9) stained with peroxidase-conjugated lectins in the presence of saponin. Cells are stained with Con A–HRP in figures 7 and 10 and with WGA–HRP in figures 8 and 9. Figure 7 illustrates the uniform intense staining of the cisternal space of the r.e.r. by Con A. The cell surface is unstained, in keeping with the very poor Con A agglutinability of these cells. Figure 8 shows that WGA is highly selective for the Golgi complex and that one face of the stack of cisternae is preferentially stained. This face is near 'G.e.r.l.-like' elements (G) and 'rigid lamellae' (r) and appears opposite the face juxtaposed to transitional elements of the r.e.r. (t.e.). Figure 9 illustrates a comparable staining gradient in a myeloma variant that no longer synthesizes Ig. Figure 10 shows that Con A stains a portion of the Golgi complex in addition to the r.e.r. The face stained is closely apposed to the r.e.r. and is hence provisionally identified as the proximal face, i.e. the face that is not stained with WGA. c, Centriole; bar = 0.5 μm.

and the studies of Rodriguez Boulan *et al.* (1978 *a*, *b*) of the luminally disposed class of rough microsomal Con A-binding glycoproteins of the rat liver.

The staining pattern with WGA–HRP is altogether different and is largely restricted to the Golgi stack, associated vesicles and multivesicular bodies. In fact, one has the impression (figure 8, plate 1) that it is only one face of the Golgi stack that is stained and that this is the face opposite the transitional elements of the r.e.r. and including 'G.e.r.l.'-like elements.

The material stained with both lectins is not simply the secretory Ig, because staining of non-producing variants of myelomas (e.g. Sp 2/0 Ag 14 of the mouse or YB 2/3.0 Ag $^{1.2.3}$ of the rat obtained from Dr Y. Argon and Dr C. Milstein) give similar staining patterns (figure 9, plate 1).

Our present biochemical analysis of WGA-binding proteins recovered from mannose gradients indicates the presence of a complex class of sialoglycoproteins ranging in molecular mass from 70 to over 200 kDa. Thus, the presence of WGA-binding sites in an anatomically distal portion of the Golgi stack correlates with the presence of terminal sugar tranferases (Roth & Berger 1982) and the above-mentioned inferences based on the use of monensin.

Finally, returning to the cells stained with Con A–HRP, there is an additional important point. In figure 7 only the r.e.r. is stained; however, when one examines the Golgi stack in such cells (figure 10, plate 2) it is clear that again one face of the stack is stained, and this time it appears to be the face closest to the r.e.r. Thus, as recently suggested on the basis of other evidence (Howell *et al.* 1978; Ito & Palade 1978; Rothman 1981), there may be r.e.r. components that are common to the Golgi stack. What is novel is that the present data suggest that they are confined to the anatomically proximal face.

Conclusion

In conclusion, the stack of cisternae of the Golgi complex shares certain features with the r.e.r. and may be subdivided by a number of means. Several altogether distinct approaches may be used to reveal this heterogeneity: perturbation by ionophores, subcellular fractionation, and ultrastructural cytochemistry. By cross-correlating the data obtained by these multiple approaches, convergent and therefore self-reinforcing conclusions should be obtained. Such a systematic mapping of the Golgi complex is still lacking but should ultimately serve as a basis for understanding the functions of this pivotal organelle.

I thank Madame M. Détraz for skilful technical assistance, Dr Y. Chicheportiche, Dr D. Hoessli and Dr J. Reigner for advice, and Dr P. Vassalli for his interest, support and suggestions. The work was supported by grant no. 3.059.0.81 from the Swiss National Science Foundation.

References

Adelman, M., Blobel, G. & Sabatini, D. 1973 Improved cell fractionation procedure for the preparation of rat liver membrane-bound ribosomes. *J. Cell Biol.* **56**, 191–205.

Bhavanandan, V. & Katlic, A. 1979 Interaction of wheat germ agglutinin with sialoglycoproteins. *J. biol. Chem.* **254**, 4000–4008.

De Duve, C. 1965 Subcellular particles. *Harvey Lect.* **59**, 49–75.

De Duve, C. 1971 Tissue fractionation, past and present. *J. Cell Biol.* **50**, 20D–55D.

Ehrenreich, J., Bergeron, J., Siekevitz, P. & Palade, G. 1973 Golgi fractions prepared from rat liver homogenates. I. *J. Cell Biol.* **59**, 45–72.

Godelaine, D., Spiro, M. & Spiro, R. 1982 Processing of the carbohydrate units of thyroglobulin. *J. biol. Chem.* **256**, 10161–10168.

Grinna, L. & Robbins, P. 1979 Glycoprotein biosynthesis. Rat liver microsomal glucosidases which process oligosaccharides. *J. biol. Chem.* **254**, 8814–8818.

Hirano, H., Parkhouse, G., Nicolson, G., Lennox, E. & Singer, S. 1972 Distribution of saccharide residues on membrane fragments from a myeloma cell homogenate. *Proc. natn. Acad. Sci. U.S.A.* **69**, 2945–2949.

Holmes, E, & O'Brien, J. 1979 Separation of glycoprotein-derived oligosaccharides by thin layer chromatography *Analyt. Biochem.* **93**, 167–170.

Howell, K., Ito, A. & Palade, G. 1978 Endoplasmic reticulum marker enzymes in Golgi fractions. What does it mean? *J. Cell Biol.* **79**, 581–589.

Howell, K. & Palade, G. 1982 Hepatic Golgi fractions resolved into membrane and content subfractions. *J. Cell Biol.* **92**, 822–832.

Ito, A. & Palade, G. 1978 Presence of NADPH-cytochrome P450 reductase in rat liver Golgi membranes. *J. Cell Biol.* **79**, 590–597.

Kornfeld, R. & Ferris, C. 1975 Interaction of immunoglobulin glycopeptides with Concanavalin A. *J. biol. Chem.* **250**, 2614–2621.

Kreibich, G. & Sabatini, D. 1974 Procedure for the selective release of content from microsomal vesicles without membrane disassembly. *Methods Enzymol.* **31**, 215–225.

Morré, D. 1971 Isolation of the Golgi apparatus. *Methods Enzymol.* **22**, 130–148.

Nakane, P. & Kawaoi, A. 1974 Peroxidase-labeled antibody. *J. Histochem. Cytochem.* **22**, 1084–1091.

Ohtsuki, I., Manzi, R., Palade, G. & Jamieson, J. 1978 Entry of macromolecular tracers into cells fixed with low concentrations of aldehydes. *Biol. cell.* **31**, 119–126.

Rodriguez Boulan, E., Kreibich, G. & Sabatini, D. 1978*a* Spatial orientation of glycoproteins in membranes of rat liver rough microsomes. I. *J. Cell Biol.* **78**, 874–893.

Rodriguez Boulan, E., Kreibich, G. & Sabatini, D. 1978*b* Spatial orientation of glycoproteins in membranes of rat liver rough microsomes. II. *J. Cell Biol.* **78**, 894–909.

Roth, J. & Berger, E. 1982 Immunocytochemical localization of galactosyltransferase in HeLa cells. *J. Cell Biol.* **93**, 223–229.

Rothman, J. 1981 The Golgi apparatus: two organelles in tendem. *Science, Wash.* **213**, 1212–1219.

Tabas, I. & Kornfeld, S. 1978 Synthesis of complex-type oligosaccharides. III. *J. biol. Chem.* **253**, 7779–7786.

Tabas, I. & Kornfeld, S. 1979 Purification and characterization of a rat liver Golgi α-D-mannosidase. *J. biol. Chem.* **254**, 11655–11663.

Tartakoff, A. 1980 The Golgi complex: crossroads for vesicular traffic. *Int. Rev. exp. Pathol.* **22**, 227–251.

Tartakoff, A. 1982 Simplifying the complex Golgi. *Trends biochem. Sci.* **7**, 174–176.

Tartakoff, A. 1983 Perturbation of the structure and function of the Golgi complex by monovalent carboxylic ionophores. *Methods Enzymol.* (*Biomembranes*) (ed. S. Fleischer & B. Fleischer). New York: Academic Press. (In the press.)

Tulsiani, D., Opheim, D. & Touster, O. 1977 Purification and characterization of α-D-mannosidase from rat liver Golgi membranes. *J. biol. Chem.* **252**, 3227–3233.

Virtanen, I., Ekblom, P. & Laurila, P. 1980 Subcellular compartmentalization of saccharide moieties in cultured normal and malignant cells. *J. Cell Biol.* **85**, 429–434.

Discussion

T. Feizi (*Division of Communicable Diseases, Clinical Research Centre, Harrow, U.K.*). Although the SP 2/0 and YB 2 myeloma cells are not secreting immunoglobulin, they are undoubtedly synthesizing and glycosylating a number of other cellular proteins. Could it be that the differing staining patterns of the proximal and distal faces of the Golgi apparatus with lectins represent reactions with 'passenger' glycoproteins rather than structural components of the Golgi membranes?

A. M. Tartakoff. No quantitative data are available allowing comparison of the relative mass of secretory and for example lysosomal glycoproteins present within the Golgi complex. Nevertheless, the wild-type myeloma cells are highly specialized toward Ig biosynthesis and one would therefore suspect that most Golgi 'passenger' glycoproteins in the wild-type are Ig. In the mutants that fail to synthesize Ig other 'passenger' glycoproteins surely predominate; however, it is not yet known to what extent they, as opposed to 'resident' glycoproteins, account for the lectin staining patterns.

Phil. Trans. R. Soc. Lond. B **300**, 185–194 (1982)
Printed in Great Britain

Synthesis and possible role of carbohydrate moieties of yeast glycoproteins

By W. Tanner, A. Haselbeck, H. Schwaiger and L. Lehle

Institut für Botanik der Universität Regensburg, 8400 Regensburg, West Germany

The pathways for protein N- and O-glycosylation in yeast cells are summarized. Evidence is presented that the terminal glucosyl residues of the dolichyl-PP-oligosaccharide intermediate are responsible for decreasing the K_m for the peptide to be N-glycosylated.

A liposomal model system is introduced that allows the study of a dolichyl phosphate (Dol-P) dependent transmembrane transport of mannosyl residues. The results obtained so far suggest that the mannosylation of Dol-P and the transmembrane translocation of Dol-P-Man are catalysed by the enzyme more or less simultaneously. However, only about 8–10 % of the enzyme molecules incorporated into the liposomes seem to carry out the 'coupled' reaction.

The glycosylation of carboxypeptidase Y is not required for this protein to reach the vacuole, its target organelle. In the presence of low concentrations of tunicamycin, however, yeast cells do stop growth. This does not seem to be due to the inhibition of secretion of glycoproteins like external invertase. It is postulated that protein glycosylation is crucial for a cell cycle event during the G1 phase.

Introduction

The participation of dolichyl phosphate activated sugars in glycoprotein synthesis has long been studied in yeast (Tanner 1969) and it is now clear that at least for N-glycosylation the pathway worked out for animal cells (Robbins *et al.* 1977; Li *et al.* 1978) also proceeds in the yeast *Saccharomyces cerevisiae* (Lehle 1981; Parodi 1981).

The reactions of the dolichol pathway are always catalysed by membrane-bound enzymes, which in all organisms are localized in the rough endoplasmic reticulum (Czichi & Lennarz 1977; Marriott & Tanner 1979, 1980).

Two major questions, however, have not yet been answered satisfactorily. Firstly, *why* are dolichyl phosphates as 'lipid intermediates' involved in this biosynthetic process? Secondly, what is the crucial general importance of protein glycosylation, evident from the fact that the dolichol pathway has obviously been preserved in a very conservative way during evolution (Lehle *et al.* 1980)? Experiments related to both these questions will mainly be discussed in this paper.

The pathway of O- and N-glycosylation in yeast

O-glycosylation

A variety of fungal glycoproteins contain short oligomannose side chains linked to serine or threonine (Sentandreu & Northcote 1969; Yen & Ballou 1973). In their synthesis dolichyl phosphorylmannose (Dol-P-Man) constitutes the mannosyl donor for the first mannosyl residue linked to protein (figure 1); subsequent mannose residues are transferred directly from

GDP-Man (Babczinski & Tanner 1973; Sharma *et al.* 1974). This reaction sequence has been described for a number of fungal cells (for references see Lehle 1981), whereas in animal cells dolichyl intermediates do not seem to participate in *O*-glycosylations (Babczinski 1980).

N-glycosylation

The statement that the dolichol pathway of *N*-glycosylation in fungal cells most probably proceeds in an identical way to that in animal cells is based on the following observations.

(*a*) The existence of dolichyl-PP-oligosaccharides containing GlcNAc, Man and Glc have been described (Parodi 1977, 1981; Lehle 1980, 1981) and the oligosaccharides synthesized *in vitro* and *in vivo* have been shown to be of the same size as the mammalian $GlcNAc_2Man_9Glc_3$ (Lehle *et al.* 1980).

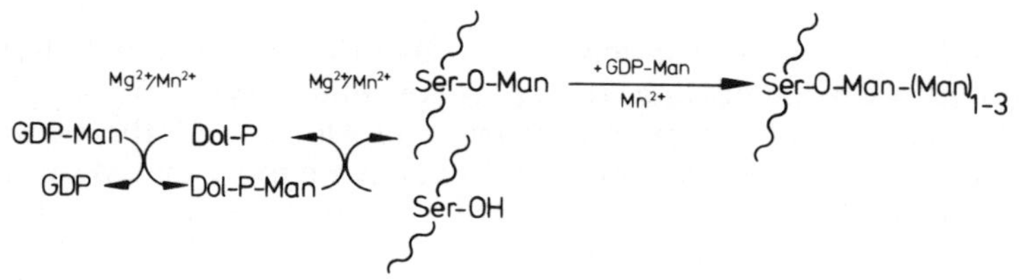

FIGURE 1. Pathway of *O*-glycosylation in yeasts.

(*b*) Dol-P-Man, GDP-Man and Dol-P-Glc that act as sugar donors for the Dol-PP-oligosaccharide seem to correspond to those of animal cells (Chapman *et al.* 1980; Parodi 1981; Lehle 1981).

(*c*) Three glucosyl residues are trimmed off the oligosaccharide after this has been transferred to the protein acceptor (Parodi 1979; Lehle 1980).

(*d*) Tunicamycin specifically blocks the GlcNAc transfer from UDP-GlcNAc to dolichyl-P (Lehle & Tanner 1976).

It is not yet clear whether the pathway ends with the glucose-trimming reactions or whether some of the mannosyl residues are also hydrolysed off, as in animal cells. Because the core structure proposed for yeast mannoproteins (Ballou 1976; Hashimoto *et al.* 1981) is not identical to the $Dol-PP-GlcNAc_2Man_9$ structure published by Li *et al.* (1978), it seems likely that 1 to 3 mannosyl residues have to be removed and subsequently added again to different hydroxyl groups.

In connection with the function of the transiently present glucose moieties it has been shown that the rate of the transfer of the oligosaccharide *en bloc* to the protein is much slower if the glucoses are missing (Turco *et al.* 1977; Spiro *et al.* 1979). By using the solubilized oligosaccharyl transferase from yeast and the hexapeptide Tyr-Asn-Leu-Thr-Ser-Val as carbohydrate-accepting molecule, this reaction was studied in greater detail (Sharma *et al.* 1981). As shown in table 1 the enzyme uses $Dol-PP-(GlcNAc)_2$ and $Dol-PP-(GlcNAc)_2Man$ almost as efficiently as $Dol-PP-(GlcNAc)_2-Man_9Glc_3$, but it does not transfer $(GlcNAc)_2Man_9$ at all to the hexapeptide. The K_m values for $Dol-PP-(GlcNAc)_2$ and $Dol-PP-(GlcNAc)_2Man_9Glc_3$ did not differ significantly (table 1). Surprisingly, however, the apparent affinity of the enzyme for the hexapeptide changed drastically with the dolichyl-PP donor molecule and was more than tenfold greater with $Dol-PP-(GlcNAc)_2-Man_9Glc_3$ than with $Dol-PP-(GlcNAc)_2$ (figure 2). This could mean that the presence of the glucosyl residues on the Dol-PP-oligosaccharide positively affects the enzyme's affinity for the protein to be glycosylated.

Does dolichol have a 'carrier' function in transmembrane transport?

As mentioned in the introduction, most of the available evidence indicates that dolichol-dependent glycosylations proceed at the endoplasmic reticulum. In fungal cells this is also true for the initial reactions of O-glycosylation (Lehle *et al.* 1977; Esmon *et al.* 1981). All these glycosylations of protein occur during the synthesis of the polypeptide chain and in all likelihood on the luminal side of the reticulum membrane (Larriba *et al.* 1976; Rothman & Lodish 1977;

TABLE 1. DONOR SUBSTRATE SPECIFICITY OF THE SOLUBILIZED
OLIGOSACCHARYL TRANSFERASE

(A radioactive count rate of 8000 min^{-1} of each of the donors was used.)

donor substrate	transfer to the hexapeptide in 15 min		apparent K_m for the donor substrate
	count min^{-1}	percentage	μM
Dol-PP-(GlcNAc)$_2$(Man)$_9$(Glc)$_3$	2667	33	0.63
Dol-PP-(GlcNAc)$_2$(Man)$_9$	51	0.6	—
Dol-PP-(GlcNAc)$_2$(Man)$_1$	2415	30	—
Dol-PP-(GlcNAc)$_2$	2089	26	1.2
Dol-PP-GlcNAc	94	1.1	—

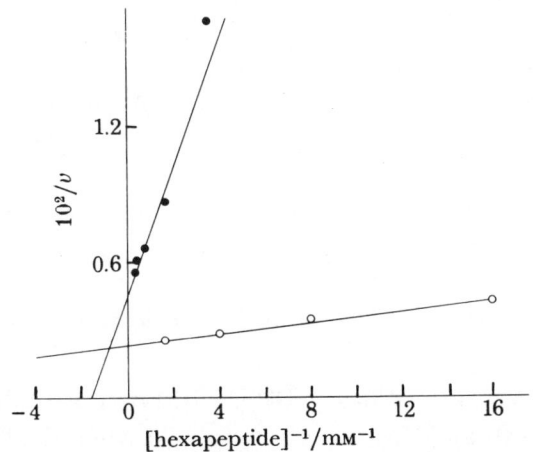

FIGURE 2. Dependence of the K_m for the hexapeptide on the glycosyl donor substrate. Standard enzyme assays were carried out for 10 min as described in Sharma *et al.* (1981) with a radioactive count rate of 8000 min^{-1} of Dol-PP-(GlcNAc)$_2$Man$_9$Glc$_3$ (○) and of Dol-PP-(GlcNAc)$_2$ (●); v is expressed as the count min^{-1} per minute of incubation.

Glabe *et al.* 1980). Because the sugar nucleotides, the ultimate precursors, are produced in the cytoplasm the problem arises how and in what form the activated sugars cross the membrane. These aspects have recently been thoroughly discussed by Hanover & Lennarz (1981) and Lennarz (this symposium). Although the dolichol-bound monosaccharides and oligosaccharides have long been suspected to represent the transmembrane transport species, this has never been demonstrated. On the contrary, neither free polyprenyl phosphates (McCloskey & Troy 1980) nor Dol-PP-(GlcNAc)$_2$ (Hanover & Lennarz 1978) move at a measurable rate across artificial or natural membranes. It has therefore been suggested that the synthesis and the transmembrane movement of the dolichol-bound saccharide might occur at the same time, i.e. the glycosyl transferase also catalysing the translocation step (Haselbeck & Tanner 1982; Hanover & Lennarz 1982).

A liposomal system for transmembrane transport

A system to test dolichyl phosphate-mediated mannosyl transfer through a lipid bilayer was set up in the following way: the mannosyltransferase of yeast cells catalysing the reversible reaction

$$\text{GDP-Man} + \text{Dol-P} \xrightleftharpoons{\text{Mg}^{2+} \text{ or } \text{Mn}^{2+}} \text{Dol-P-Man} + \text{GDP}$$

was solubilized and purified approximately a thousandfold (Babczinski 1980; Haselbeck & Tanner 1982). This enzyme was incorporated into soybean lecithin liposomes containing

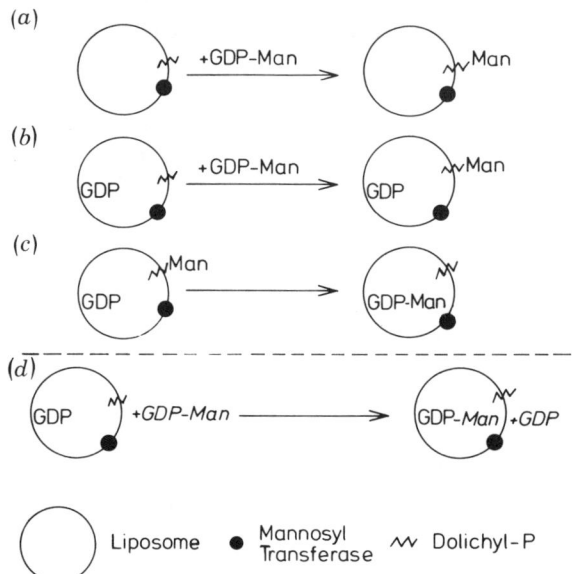

Figure 3. Dolichyl phosphate-mediated mannosyl transfer into liposomes: schematic reaction sequence.

Dol-P (Dol-P:lecithin ratio 1:17). When GDP-[^{14}C]Man was added, the liposomal Dol-P was mannosylated (figure 3a), and this was also observed when the liposomes were preloaded with 7 mM GDP (figure 3b). However, only in the latter case did the liposomes, which were separated from the excess external GDP-[^{14}C]Man by a Sepharose 4B column, also contain a water-soluble radioactive compound (figure 4). This was identified as GDP-[^{14}C]Man (Haselbeck & Tanner 1982) and the result has therefore been interpreted as a transmembrane transfer of a mannosyl residue from external GDP-Man to internal GDP (figure 3d). That Dol-P is indeed required in the reaction sequence shown in figure 3d, and that Dol-P-Man is therefore most probably a transport intermediate, as shown in figure 3c, is indicated by the data of table 2.

From the results so far presented it is not possible to distinguish between a mannosylation of liposomal Dol-P and a subsequent transmembrane movement of Dol-P-Man ('flip-flop') and the second possibility of a simultaneous mannosylation and translocation; in the latter case both events would be catalysed by the enzyme. Two observations are in favour of the second possibility.

(a) The time course of formation of internal GDP-Man and liposomal Dol-P-Man in liposomes either preloaded or not preloaded with GDP is shown in figure 5. In GDP-preloaded liposomes the internal GDP-[14C]Man formed always amounts to about 8 % of the total liposomal Dol-P-[14C]Man synthesized. This can be interpreted to mean that only a small fraction of the enzyme carries out the whole reaction, i.e. mannosylation plus translocation,

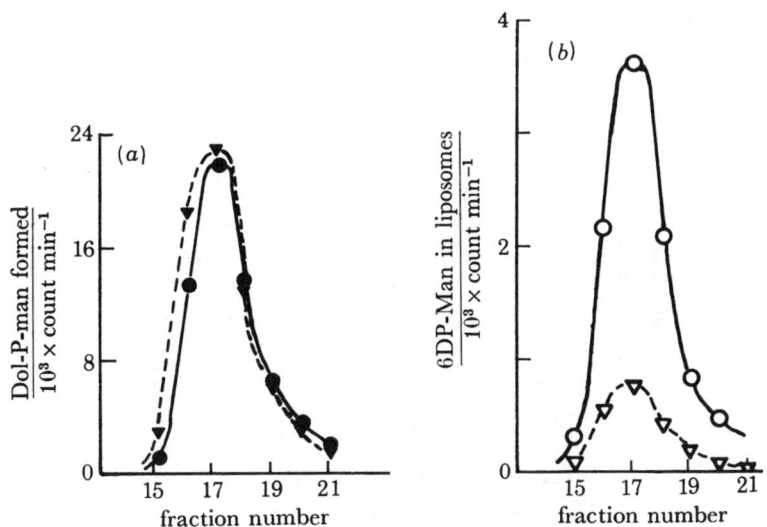

FIGURE 4. Radioactivity from GDP-[14C]Man incorporated into (a) the lipid fraction and (b) the water-soluble fraction of Dol-P-containing liposomes. The liposomes were separated from the incubation medium on a Sepharose 4B column; they appeared in the void volume. —, Liposomes with GDP; – – –, liposomes without GDP. For other experimental details see Haselbeck & Tanner (1982).

TABLE 2. MANNOSYL TRANSFER FROM EXTERNAL GDP-[14C]MAN TO INTERNAL GDP:
DEPENDENCE ON DOL-P AND ON MANNOSYL TRANSFERASE

| | radioactive count min^{-1} in: | |
conditions	Dol-P-Man	GDP-Man in liposomes
complete liposomes	28 911	1491
minus Dol-P	222	36
minus enzyme	24	20
complete liposomes with amphomycin (75 μg)	11	109

whereas most of it only catalyses the mannosylation step. In case of a rate limiting 'flip-flop' of Dol-P-[14C]Man, an initial lag in the formation of internal GDP-[14C]Man and a subsequent steady increase in its percentage of the total liposomal radioactivity would have been expected. That most of the liposomal Dol-P-Man formed is indeed oriented with its mannosyl residue to the outside is inferred from the observation that approximately 75 % of its radioactivity is rapidly lost when 1 mM GDP is added to the external medium (Haselbeck 1982).

(b) When liposomes were prepared with polyprenyl phosphates of differing chain lengths, the rate of formation of internal GDP-[14C]Man was more or less independent of the kind of polyprenyl phosphate used, whereas the rate of polyprenyl-P-Man formation differed significantly (table 3). For a Dol-P-Man 'flip-flop', a translocation rate (measured as internal GDP-Man formed with time) proportional to the radioactivity present in Dol-P-Man would again have been expected.

[81]

Comparison of activities in vitro *and* in vivo

Assuming, then, that the mannosyl transferase catalyses simultaneously both the mannosylation of Dol-P as well as the transmembrane translocation of Dol-P-Man – whereas in the liposomal system both these steps seem to be 'uncoupled' to a large extent – the 'internalized' mannosyl residue might subsequently undergo two types of reactions. Either it gets transferred directly from membrane-bound Dol-P-Man to the intraluminally growing Dol-PP-(GlcNAc)$_2$-Man$_5$ to form Dol-PP-(GlcNAc)$_2$Man$_9$ or it gets transferred back to GDP present in the e.r.

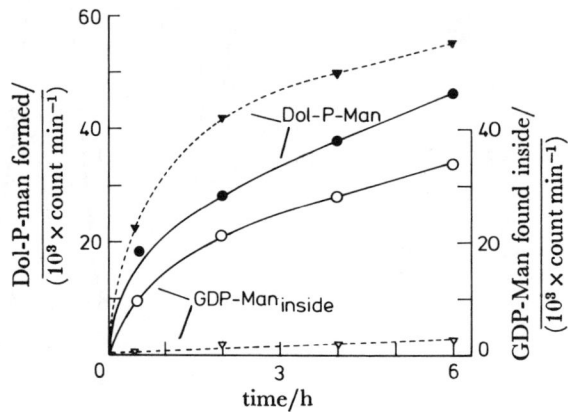

FIGURE 5. Time course of [^{14}C]mannose transfer from GDP-[^{14}C]Man to dolichyl phosphate and to internal GDP. —, Liposomes with GDP; – – –, liposomes without GDP.

TABLE 3. DEPENDENCE OF MANNOSYL TRANSFER ON POLYPRENYL CHAIN LENGTH

	Dol-P-Man formed	mannosyl residue transported into liposome	
	(count min^{-1})	(count min^{-1})	percentage of Dol-P-Man
C$_{35}$–Dol-P	170 294	4100	2.4
C$_{60}$–Dol-P	79 447	3382	4.2
C$_{80-100}$–Dol-P	40 225	4134	10.3

lumen analogous to the situation in the liposomal system. The internal GDP-Man could then serve as mannosyl donor for all those reactions that have been shown *in vitro* to require GDP-Man directly. In the yeast cell the latter reaction would be involved in the formation of the *O*-glycosidic side chains (except for the first mannosyl residue) as well as in the extension of the outer chains of mannoproteins (Parodi 1981; Lehle 1981), which can amount to more than two-thirds of the carbohydrate content of yeast mannoproteins (Ballou 1976). Thus if all mannose residues of yeast mannoproteins require the formation of Dol-P-Man for their transmembrane transport, the mannosyl transferase activity *in vivo* would have to be considerably greater than that required for reactions directly dependent on Dol-P-Man. For a mannose content of 10 % of yeast dry mass (Sentandreu & Northcote 1968) and a generation time of 2 h, a Dol-P-Man forming activity of 310 μmol h^{-1} g^{-1} yeast dry mass would be required. This has to be compared with a maximal activity measured *in vitro* of 165 μmol h^{-1} g^{-1} yeast dry mass at saturating GDP-Man and optimal Dol-P concentrations. Although substrate

[82]

saturation possible does not exist *in vivo*, when all the possible fallacies of work *in vitro* are considered, the two values do not differ to an extent that would exclude Dol-P-Man as a general transmembrane transport vehicle for mannose in *S. cerevisiae*.

Why are yeast proteins glycosylated anyway?

In recent years an increasing number of examples concerning the functional importance of carbohydrate moieties of glycoproteins became known. Certainly among the best documented are those of receptor-mediated pinocytosis (Neufeld & Ashwell 1980). Since yeast cells contain

TABLE 4. DISTRIBUTION OF GLYCOSYLATED AND NON-GLYCOSYLATED
(CARBOHYDRATE-FREE) CARBOXYPEPTIDASE Y IN CELL FRACTIONS

	carboxypeptidase Y		carbohydrate-free carboxypeptidase Y	
	count min^{-1}	percentage	count min^{-1}	percentage
cytosol	768	18.3	612	16.9
vacuoles	2598	62.1	2126	58.7
pellet	894	21.4	854	23.4

a lysosome-like vacuole, with a number of enzymically active glycoproteins it was natural to speculate that in this organism too the carbohydrate moieties are responsible for tragefng the enzymes to the vacuole. Carboxypeptidase Y has been studied in more detail. It was shown that the enzyme contains four asparagine-linked oligosaccharides (Trimble & Maley 1977; Hasilik & Tanner 1978b) and that phosphomonoester and phosphodiester groups are indeed present within a larger oligomannose moiety (Hashimoto *et al.* 1981; Schwaiger *et al.* 1982). In the presence of tunicamycin, carbohydrate-free carboxypeptidase Y is synthesized, although in a reduced amount (Hasilik & Tanner 1978b). Is this carbohydrate-free protein still transported into the vacuole? Cells treated with tunicamycin contained approximately equal amounts of [^{14}C]phenylalanine-labelled intact carboxypeptidase Y and the carbohydrate-free form (Schwaiger *et al.* 1982), and both forms were found to be present in vacuoles isolated from these cells to the same extent (table 4). Thus, obviously neither mannose 6-phosphate residues nor any other part of the carboxypeptidase Y carbohydrate moiety is required as sorting signal for the protein to reach the vacuole. Evidence emerges that also in mammalian cells an additional sorting signal besides mannose 6-phosphate has to exist (Owada & Neufeld 1982; Waheed *et al.* 1982; Jessup & Dean 1982).

Because yeast carboxypeptidase, like other lysosomal enzymes, is synthesized via a 'pro' form with an extension of the peptide chain (Hasilik & Tanner 1976, 1978a), the sorting information could of course be located in this extra peptide piece (M_r 6000). A mutant deficient in carboxypeptidase Y processing (Hemmings *et al.* 1981) still transports the procarboxypeptidase Y into the vacuole (Schwaiger, unpublished). Thus the conversion of the pro form into the final carboxypeptidase Y form is not required for the protein to be put into the vacuole. This does not, however, rule out the possible role of the extra peptide piece as a target signal.

Nevertheless, the functional importance of the carbohydrate moiety of carboxypeptidase Y is again completely open and so is the question posed in the introduction about a crucial role of protein glycosylation in general. For yeast cells, and for many others, it has been shown that the secretion of proteins is inhibited when glycosylation is prevented, for example with tunicamycin (Kuo & Lampen 1974). The important question in this context is, however, whether yeast cells stop growing, when *N*-glycosylation is inhibited by tunicamycin. So far this has not been observed when reasonably low inhibitor concentrations have been used (Hasilik & Tanner 1976); these cells were in the late logarithmic phase. Their growth is, however, completely inhibited with tunicamycin at 4–10 µg ml⁻¹ when low-density logarithmically growing cultures

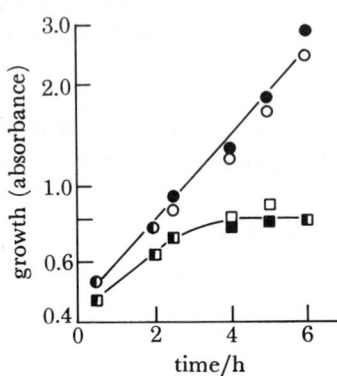

FIGURE 6. Growth inhibition of yeast cells (*S. cerevisiae* LBC H 1022 aα) by tunicamycin (8 µg ml⁻¹). Open symbols, glucose; filled symbols, sucrose; circles, without tunicamycin; squares, with tunicamycin.

are used (figure 6). Since yeast cells growing on sucrose require the presence of external invertase for growth, the effect of tunicamycin under these conditions was expected. The inhibition is identical, however, when the cells are grown on glucose (figure 6). This indicates that the complete cessation of growth after about 3 h is not due to the interference of tunicamycin with the synthesis and secretion of invertase. An analysis of the inhibited cells showed that they all stopped growth in the G1 phase of the cell cycle (Arnold & Tanner 1982). It is therefore tempting to speculate that protein glycosylation is crucial for a cell cycle event during the G1 and before the S phase. This hypothesis will have to be carefully checked in the future.

Thanks are due to Cornelia Weber for carrying out the experiment whose results are shown in figure 6 and to Dr T. Chojnacki for supplying polyprenyl phosphates of various chain lengths. The original work from this laboratory mentioned in the text has been supported by the Deutsche Forschungsgemeinschaft (SFB 43).

REFERENCES

Arnold, E. & Tanner, W. 1982 An obligatory role of protein glycosylation in the life cycle of yeast cells. *FEBS Lett.* (Submitted.)
Babczinski, P. 1980 Evidence against the participation of lipid intermediates in the *in vitro* biosynthesis of serine (threonine)-*N*-acetyl-D-galactosamine linkages in submaxillary mucin. *FEBS Lett.* **117**, 207–211.
Babczinski, P., Haselbeck, A. & Tanner, W. 1980 Yeast mannosyl transferases requiring dolichyl phosphate and dolichyl phosphate mannose as substrate. Partial purification and characterization of the solubilized enzyme. *Eur. J. Biochem.* **205**, 509–515.
Babczinski, P. & Tanner, W. 1973 Involvement of dolichol monophosphate in the formation of specific mannosyl linkages in yeast glycoproteins. *Biochem. biophys. Res. Commun.* **54**, 1119–1129.
Ballou, C. E. 1976 Structure and biosynthesis of the mannan component of the yeast cell envelope. *Arch. microb. Physiol.* **14**, 93–158.
Chapman, A., Fujimoto, K. & Kornfeld, S. 1980 The primary glycosylation defect in class E Thy-1-negative mutant mouse lymphoma cells is an inability to synthesize dolichol-P-mannose. *J. biol. Chem.* **255**, 4441–4446.

Czichi, U. & Lennarz, W. J. 1977 Localization of the enzyme system for glycosylation of proteins via the lipid-linked pathway in rough endoplasmic reticulum. *J. biol. Chem.* **252**, 7901–7904.

Esmon, B., Novick, P. & Schekman, R. 1981 Compartmentalized assembly of oligosaccharides on exported glycoproteins in yeast. *Cell* **25**, 451–460.

Glabe, C. G., Hanover, J. A. & Lennarz, W. J. 1980 Glycosylation of ovalbumin nascent chains. *J. biol. Chem.* **255**, 9236–9242.

Hanover, J. A. & Lennarz, W. J. 1978 The topological orientation of N,N'-diacetylchitobiosyl-pyrophosphoryl-dolichol in artificial and natural membranes. *J. biol. Chem.* **254**, 9237–9246.

Hanover, J. A. & Lennarz, W. J. 1981 Transmembrane assembly of membrane and secretory glycoproteins. *Arch. Biochem. Biophys.* **211**, 1–19.

Hanover, J. A. & Lennarz, W. J. 1982 Transmembrane assembly of N-linked glycoproteins. *J. biol. Chem.* **257**, 2787–2794.

Haselbeck, A. 1982 Doctoral dissertation, Universität Regensburg.

Haselbeck, A. & Tanner, W. 1982 Dolichyl phosphate-mediated mannosyl transfer through liposomal membranes. *Proc. natn. Acad. Sci. U.S.A.* **79**, 1520–1524.

Hashimoto, C., Cohen, R. E., Zhang, W.-J. & Ballou, C. E. 1981 Carbohydrate chains on yeast carboxypeptidase Y are phosphorylated. *Proc. natn. Acad. Sci. U.S.A.* **78**, 2244–2248.

Hasilik, A. & Tanner, W. 1976 Biosynthesis of carboxypeptidase Y in yeast. Evidence for a precursor form of the glycoprotein. *Biochem. biophys. Res. Commun.* **72**, 1430–1436.

Hasilik, A. & Tanner, W. 1978a Biosynthesis of the vacuolar yeast glycoprotein carboxypeptidase Y. Conversion of precursor into the enzyme. *Eur. J. Biochem.* **85**, 599–608.

Hasilik, A. & Tanner, W. 1978b Carbohydrate moiety of carboxypeptidase Y and perturbation of its biosynthesis. *Eur. J. Biochem.* **91**, 567–575.

Hemmings, B. A., Zubenko, G. S., Hasilik, A. & Jones, E. W. 1981 Mutant defective in processing of an enzyme located in the lysosome-like vacuole of *S. cerevisiae. Proc. natn. Acad. Sci. U.S.A.* **78**, 435–439.

Jessup, W. & Dean, R. T. 1982 Secretion by mononuclear phagocytes of lysosomal hydrolases bearing ligands for the mannose-6-phosphate receptor system of fibroblasts: evidence for a second mechanism of spontaneous secretion? *Biochem. biophys. Res. Commun.* **105**, 922–927.

Kuo, S.-C. & Lampen, J. O. 1974 Tunicamycin – an inhibitor of yeast glycoprotein synthesis. *Biochem. biophys. Res. Commun.* **58**, 287–295.

Larriba, G., Elorza, M. V., Villanueva, J. R. & Sentandreu, R. 1976 Participation of dolichol phosphomannose in the glycosylation of yeast wall mannoproteins at the polysomal level. *FEBS Lett.* **71**, 316–320.

Lehle, L. 1980 Biosynthesis of the core region of yeast mannoproteins. *Eur. J. Biochem.* **109**, 589–601.

Lehle, L. 1981 Biosynthesis of mannoproteins in fungi. In *Encyclopedia of plant physiology* (new series), vol. 13 B (ed. W. Tanner & F. A. Loewus), pp. 459–483. Berlin, Heidelberg and New York: Springer-Verlag.

Lehle, L., Bauer, F. & Tanner, W. 1977 The formation of glycosidic bonds in yeast glycoproteins. Intracellular localisation of the reactions. *Arch. Microbiol.* **114**, 77–81.

Lehle, L., Schulz, I. & Tanner, W. 1980 Dolichyl phosphate linked sugars as intermediates in the synthesis of yeast mannoproteins: an *in vivo* study. *Arch. Microbiol.* **127**, 231–237.

Lehle, L. & Tanner, W. 1976 The specific site of tunicamycin inhibition in the formation of dolichol-bound N-acetyl-glucosamine derivatives. *FEBS Lett.* **71**, 167–170.

Li, E., Tabas, I. & Kornfeld, S. 1978 The synthesis of complex-type oligosaccharides. *J. biol. Chem.* **253**, 7762–7770.

McCloskey, M. A. & Troy, F. A. 1980 Paramagnetic isoprenoid carrier lipids. 2. Dispersion and dynamics in lipid membranes. *Biochemistry, Wash.* **19**, 2061–2066.

Marriott, K. M. & Tanner, W. 1979 Dolichylphosphate dependent glycosyl transfer reactions in the endoplasmic reticulum of castor bean endosperm. *Pl. Physiol.* **64**, 445–449.

Marriott, M. & Tanner, W. 1980 Localization of dolichyl phosphate- and pyrophosphate-dependent glycosyl transfer reactions in *S. cerevisiae. J. Bact.* **139**, 565–572.

Neufeld, E. & Ashwell, G. 1980 Carbohydrate recognition systems for receptor mediated pinocytosis. In *The biochemistry of glycoproteins and proteoglycans* (ed. W. J. Lennarz), pp. 241–266. New York and London: Plenum Press.

Owada, M. & Neufeld, E. F. 1982 Is there a mechanism for introducing acid hydrolases into liver lysosomes that is independent of mannose 6-phosphate recognition? *Biochem. biophys. Res. Commun.* **105**, 814–820.

Parodi, A. J. 1977 Synthesis of glycosyl-dolichol derivatives in baker's yeast and their role in protein glycosylation. *Eur. J. Biochem.* **83**, 253–259.

Parodi, A. J. 1979 Biosynthesis of yeast glycoproteins. *J. biol. Chem.* **254**, 10051–10060.

Parodi, A. J. 1981 Biosynthetic mechanisms for cell envelope polysaccharides. In *Yeast cell envelopes: biochemistry, biophysics and ultrastructure*, vol. 2 (ed. W. N. Arnold), pp. 47–64. Boca Raton: CRC Press, Inc.

Robbins, P. W., Hubbard, S. C., Turco, S. J. & Wirth, D. F. 1977 Proposal for a common oligosaccharide intermediate in the synthesis of membrane glycoproteins. *Cell* **12**, 893–900.

Rothman, J. E. & Lodish, H. F. 1977 Synchronized transmembrane insertion and glycosylation of a nascent membrane protein. *Nature, Lond.* **269**, 775–780.

194 W. TANNER AND OTHERS

Schwaiger, H., Hasilik, A., von Figura, K., Wiemken, A. & Tanner, W. 1982 Carbohydrate-free carboxy-peptidase Y is transferred into the lysosome-like yeast vacuole. *Biochem. biophys. Res. Commun.* **104**, 950–956.

Sentandreu, R. & Northcote, D. H. 1968 The structure of a glycopeptide isolated from the yeast cell wall. *Biochem. J.* **109**, 419–432.

Sentandreu, R. & Northcote, D. H. 1969 The characterization of oligosaccharide attached to threonine and serine in mannan glycopeptides obtained from the cell wall of yeast. *Carbohydr. Res.* **10**, 584–585.

Sharma, C. B., Babczinski, P., Lehle, L. & Tanner, W. 1974 The role of dolichol monophosphate in glyco-protein biosynthesis in *S. cerevisiae*. *Eur. J. Biochem.* **46**, 35–41.

Sharma, C. B., Lehle, L. & Tanner, W. 1981 *N*-Glycosylation of yeast proteins. Characterization of the solu-bilized oligosaccharyl transferase. *Eur. J. Biochem.* **116**, 101–108.

Spiro, M. J., Spiro, R. G. & Bhoyroo, V. D. 1979 Glycosylation of proteins by oligosaccharide lipids. *J. biol. Chem.* **254**, 7668–7674.

Tanner, W. 1969 A lipid intermediate in mannose biosynthesis in yeast. *Biochem. biophys. Res. Commun.* **35**, 144–150.

Trimble, R. B. & Maley, F. 1977 The use of endo-β-*N*-acetyl-glucosaminidase H in characterizing the structure and function of glycoproteins. *Biochem. biophys. Res. Commun.* **78**, 935–944.

Turco, S. J., Stetson, B. & Robbins, P. W. 1977 Comparative rates of transfer of lipid-linked oligosaccharide to endogenous glycoprotein acceptors *in vitro*. *Proc. natn. Acad. Sci. U.S.A.* **74**, 4411–4414.

Waheed, A., Pohlmann, R., Hasilik, A., von Figura, K., van Elsen, A. & Leroy, J. G. 1982 Deficiency of UDP-*N*-acetylglucosamine: lysosomal enzyme *N*-acetylglucosamine-1-phosphotransferase in organs of I-cell patients. *Biochem. biophys. Res. Commun.* **105**, 1052–1058.

Yen, P. H. & Ballou, C. E. 1973 Composition of a specific intercellular agglutination factor. *J. biol. Chem.* **248**, 8316–8318.

Phil. Trans. R. Soc. Lond. B **300**, 195–206 (1982)

Printed in Great Britain

The synthesis and transport of some plant glycoproteins

By D. H. Northcote, F.R.S.

Department of Biochemistry, University of Cambridge,
Tennis Court Road, Cambridge CB2 1QW, U.K.

Maize root slime is secreted by the outer root-cap cells. It is a complex of at least three polysaccharides, two of which are acidic and contain a high proportion of fucose. The polysaccharides are assembled and secreted by the endoplasmic reticulum and dictyosomes of the cells. It has been found that the synthesis of fucose-containing oligo-saccharides takes place by an assembly of at least nine sugar residues on a polyprenyl diphosphate acceptor and this occurs at the endoplasmic reticulum. In addition, a glycoprotein carrying a large molecular mass carbohydrate portion containing fucose is synthesized mainly within the Golgi apparatus, although synthesis may be initiated in the endoplasmic reticulum. This glycoprotein could be an intermediate in the synthesis of slime polysaccharide, and the carbohydrate moiety of the glycoprotein may be assembled by transglycosylations from the polyprenyl diphosphate oligosaccharides.

Introduction

The outer root-cap cells of many plants secrete a slime that serves to keep the root moist and to lubricate and ease the passage of the growing root in the soil. These cells occur at the first few millimetres of the root tip in the root cap and in addition to the secretion of slime they are themselves continuously sloughed off and autolysed so that the root is covered not only with the secreted material but the products of the autolysed cells (Barlow 1975; Rougier 1981).

We have studied the structure and the mechanism of synthesis and secretion of the slime produced by the maize root tip. This material can be obtained as a droplet at the tip of sterile roots grown so that the tissue is suspended in air, and it can be washed off the root or collected on filter paper, or it can be collected from a mass of rootlets produced on a callus grown as a tissue culture on solid medium. It is also secreted into the liquid medium of a suspension culture of maize cells or into a liquid culture of maize primary roots (figure 1) (Wright & Northcote 1974). The slime collected direct from the intact root contains approximately 30 % protein and the rest of the dry matter of the material is carbohydrate, mainly polysaccharide (Wright & Northcote 1975).

Analysis and general structure of the slime

The polysaccharides in the material consist of at least three polymers that can be separated by electrophoresis on glass fibre paper (Wright & Northcote 1974). These are a neutral polymer, a weakly acidic polymer and an acidic polymer. The sugar composition of these polymers varies with the preparation, but the general nature of the materials as shown by measurement of their radioactive composition by feeding the roots [14C]glucose is given in table 1.

The neutral polymer is mainly made up of glucose residues and consists of a β 1–4 glucan, whereas the acidic polymers contain fucose, arabinose, galactose and galacturonic acid residues with minor amounts of xylose, glucose and mannose. It is the variation in the uronic acid

FIGURE 1. Maize roots grown in liquid culture. The roots originated on a solid callus derived from a shoot of a maize seedling. The callus was maintained on Hellers medium with coconut milk (20 % by volume), sucrose (20 g l⁻¹) and 2,4-dichlorophenoxyacetic acid (6 mg l⁻¹) (Wright & Northcote 1974). The root culture was grown in half-strength Murashige & Skoog's medium (Murashige & Skoog 1962) with the addition of casamino acids(1 g l⁻¹), sucrose (25 g l⁻¹) and 3-naphthylacetic acid (2.5 mg l⁻¹).

TABLE 1. COMPONENTS OF MAIZE-ROOT SLIME

(The analyses given in this table relate to the relative proportions of radioactive sugars in the components. The tissue had been exposed to radioactive glucose for 2.5 h and during this length of time the slime became extensively radioactive and the cytoplasmic pools of slime precursors became saturated with radioactivity. It is therefore probable that, although the analysis does not indicate an accurate chemical composition, it does reflect the differences between the components. The radioactive composition of the unfractionated material is very similar to that of the chemical composition of the slime (Harris & Northcote 1970).)

| | radioactivity (%) | | | |
sugar	unfractionated material	neutral component	weakly acidic component	strongly acidic component
uronic acid	16	—	19	57
galactose	21	—	31	18
glucose	23	91	4	0
mannose	1	—	4	3
arabinose	11	—	12	16
xylose	8	—	6	3
fucose	20	—	23	3

content that causes the separation of the weakly acidic from the acidic fractions of the slime polysaccharides. The overall composition of the slime compared with the composition of the separated fractions is shown in table 1.

It is probable that the complex is made up of a central core of fairly rigid β 1–4 glucan chains that are held together and are kept extended by intramolecular and intermolecular hydrogen bonds, not dissimilar from the chains in cellulose microfibrils but shorter. This central core is surrounded by a hydrophilic coat of the acidic polymers (Wright & Northcote 1976; Grant et al.

1969). The acidic polymers do not seem to have continuous chains of galacturonic acid units as in pectin but to consist of either mixed chains of glucose and uronic acid residues carrying side chains of the neutral sugars, or chains of glucose units with the acidic and neutral sugar residues carried as side chains (Wright & Northcote 1975; Wright et al. 1976).

The large amount of fucose in the weakly acidic polymer is unusual and allows the metabolism of the polymer, which is only produced by the differentiated cells of the root tip, to be uniquely followed and it is this polymer that will be subsequently discussed in this work. The presence of such large quantities of fucose makes it possible to study the location and synthesis of this polysaccharide, because fucose applied to the roots is taken up and is not metabolized to carbon dioxide and water but is all incorporated into the polymers without breakdown or conversion into other sugars (Wright & Northcote 1976; Kirby & Roberts 1971; Paull & Jones 1975).

LOCATION OF SYNTHESIS

Radioautographic studies on the cells of wheat roots incubated with radioactive glucose (Northcote & Pickett-Heaps 1966) or in addition, in maize roots, radioactive fucose, have indicated that the slime is produced by the outer cap cells (Kirby & Roberts 1971; Rougier 1981). The endomembrane system is involved and the radioactive slime can be shown to be secreted across the plasma membrane by vesicle fusion and an exopinocytotic mechanism. The vesicles are derived from the Golgi apparatus and the radioautograms indicate that a concentration of the slime and possibly its active synthesis occur in this organelle (Rougier 1976; Paull & Jones 1976).

The work has been extended by membrane fractionation of the cells and the analysis of radioactive material located in characterized membrane fractions after feeding the roots a radioactive precursor such as [U-^{14}C]glucose (Harris & Northcote 1971; Bowles & Northcote 1972). This type of study has provided evidence for the function of both the endoplasmic reticulum and the Golgi apparatus for the production and transport of the slime from the cytoplasm. In these experiments maize roots were incubated with radioactive glucose and the membranes were isolated on a discontinuous sucrose gradient after breakage of the cells. The cells were fractionated into wall material and fractions enriched in microsomes, dictyosomes, mitochondria, smooth membranes and soluble polymers. The membranes were characterized by observations with the electron microscope and by their enzyme content. The composition of the radioactive polymers that they carried was determined (Bowles & Northcote 1972; Bowles & Northcote 1976).

Over the period of the incubation with the radioactive glucose, the type of polymer that is being incorporated into the cell wall or the slime can be determined. This varies with the position of the cell in the root. The cells that are differentiated into outer root-cap cells produce slime containing fucose. Further up the root the cells become differentiated to form vascular tissue and the walls are secondary-thickened, and there is an increase in the amount of polymers containing xylose that are secreted into the wall. At the root-tip, just behind the cap cells, the meristematic tissue is actively dividing and pectin that contains arabinose is laid down in relatively large amounts into the wall. These differences in polymer formation by the various differentiated cells are reflected in the type of polymer found in the membrane system (Bowles & Northcote 1972). The synthetic systems for oligosaccharide formation found at the membrane change, and these are the control steps whereby one aspect of the differentiation is manifested

(Northcote 1982a). The information that is gained from such studies clearly shows that polymers that contain fucose are present both at the endoplasmic reticulum and the Golgi apparatus. It shows that the membranes at which these polymers are located and almost certainly formed occurs only in those cells in the first 1–2 mm of the root tip at the root cap (Bowles & Northcote 1972).

TABLE 2. THE AMOUNTS OF ROOT-SLIME POLYSACCHARIDE COMPONENTS IN THE DICTYOSOME AND ROUGH ENDOPLASMIC RETICULUM FRACTIONS, AND THE RATE OF INCREASE IN ROOT SLIME OF MAIZE ROOTS

| | amount of material/pmol | | rate of increase into slime polysaccharide |
sugar	rough endoplasmic reticulum	Golgi apparatus	pmol min^{-1}
galactose	5.7	0.3	0.87
glucose	6.0	0.3	0.91
mannose	1.6	0.09	0.25
arabinose	4.0	0.2	0.62
xylose	1.1	0.06	0.17
fucose	8.7	0.5	1.33
total	**27.1**	**1.5**	**4.15**

It is also apparent from the observations on the membranes, both *in situ* and by analysis of their contents, that the slime complex as a whole is synthesized and secreted probably as an organized unit with the core glucan surrounded by the acidic polymers. The secretion of skeletal scales in the alga *Chrysochromulina chiton* clearly illustrates that the membrane system of cells can secrete such polymers organized into a complex composite structure with cellulose-like microfibrils embedded in a matrix of polysaccharides and glycoprotein (Allen & Northcote 1975).

In the maize root it can be shown that, during an incubation period of the roots with radioactive glucose over a period of 30–40 min, the internal pools of polysaccharide-like polymers that are being formed and that are being continuously secreted by the cells, become saturated with radioactivity. It is thus possible to calculate the sizes of these pools and to estimate their rate of turnover to produce the amount of polysaccharide secreted (Bowles & Northcote 1974). Since the approximate composition of the slime is known and since this is the only polysaccharide that contains fucose, the rate of polysaccharide secretion can be subdivided into that for wall polysaccharide secretion and that for slime polysaccharide secretion. The rates of polysaccharide secretion and the estimated pool sizes are shown in tables 2 and 3. The values are suggested as an indication of the magnitude of the turnover because the membrane fractions from which the calculations were made were not pure but were enriched with the particular membranes. However, it can be seen that the rate of secretion of the slime by the comparatively few cells of the root cap is very high and that one aspect of their differentiation is the rapid turnover rate of polysaccharide synthesis and, connected with this, the rate of vesicle fusion at the membrane. The rate of turnover of the membranes in the root cap for slime production is approximately eight times as fast as that for the production of wall polysaccharides. It has been assumed in these calculations that the polymer material is secreted in the sequence from endoplasmic reticulum to Golgi apparatus to the extracellular space, and that there is no direct contribution from the endoplasmic reticulum to the outside of the cell.

TABLE 3. THE AMOUNTS OF CELL-WALL POLYSACCHARIDE COMPONENTS IN THE DICTYOSOME AND ROUGH ENDOPLASMIC RETICULUM FRACTIONS, AND THE RATE OF INCREASE IN CELL-WALL POLYSACCHARIDES OF MAIZE ROOTS

| | amount of material/pmol | | rate of increase into cell-wall polysaccharide |
| | rough endoplasmic | Golgi | |
sugar	reticulum	apparatus	pmol min⁻¹
galactose	38.7	1.8	1.1
glucose	19.2	0.56	0.8
mannose	2.8	0.01	0.0
arabinose	58.9	3.0	0.8
xylose	117.2	6.7	2.1
fucose	0	0	0
total	**236.8**	**12.1**	**4.8**

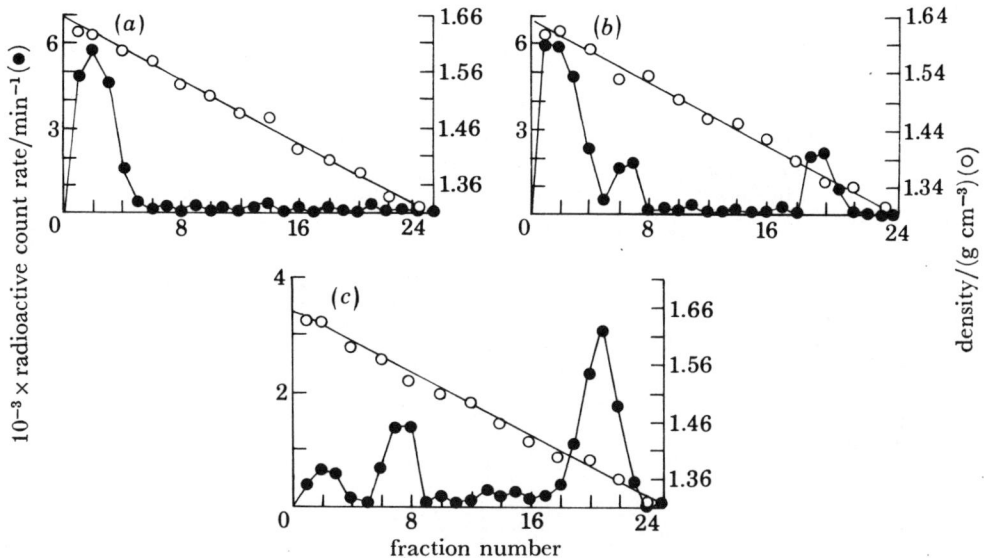

FIGURE 2. Distribution of radioactivity (●) in CsCl density gradients after centrifuging slime extracted from the intact roots (a) an aqueous extract of the broken cells of root tips (b) and an aqueous extract of sonicated membranes (c) prepared from the broken cells of root tips. The roots were incubated with L[1-³H]fucose for 1 h at 20 °C in the dark before the slime and the aqueous extracts were prepared (Green & Northcote 1978).

INTERMEDIATES DURING THE SYNTHESIS OF THE SLIME POLYSACCHARIDE

Glycoproteins

Polymers can be extracted from the isolated membrane systems of the cells by breaking the membranes by homogenization and sonication to obtain water-soluble material. These solutions can be fractionated on a gradient of caesium chloride solution (Paull & Jones 1976) to give bands of polymers at different densities. In this way polysaccharide (density 1.6–2.0 g cm⁻³) can be separated from protein density (1.3 g cm⁻³) and glycoproteins (density between 1.6 and 1.3 g cm⁻³) (Green & Northcote 1978) (figure 2).

The slime taken directly from the intact root tips contained only polysaccharide and protein; no glycoprotein could be detected. However, within the membrane system of the cytoplasm there were three polymers that incorporated radioactive fucose, two glycoproteins at densities of 1.55 and 1.37 g cm^{-3}, and a polysaccharide. There was in addition some free protein in the solutions. All the polymers could be made radioactive when the roots were incubated with radioactive glucose, whereas radioactive leucine was incorporated into those polymers that contained protein. The glycoproteins and polysaccharides were isolated from the gradient and

TABLE 4. RELATIVE AMOUNTS OF RADIOACTIVITY INCORPORATED FROM D-[U-^{14}C]GLUCOSE INTO THE POLYSACCHARIDE COMPONENTS OF DIFFERENT FRACTIONS OBTAINED FROM A CsCl DENSITY GRADIENT

	radioactivity of the isolated fractions (percentage of total)		
sugar	density 1.63 g cm^{-3}	density 1.55 g cm^{-3}	density 1.37 g cm^{-3}
uronic acid	15.6	15.0	12.5
galactose	21.4	17.5	20.2
glucose	23.2	27.5	22.8
mannose	1.0	0.8	0.7
arabinose	10.8	10.4	11.5
xylose	8.3	13.6	7.3
fucose	19.7	15.2	25.0
total radioactive count rate/min^{-1}	15 300	8 500	12 200

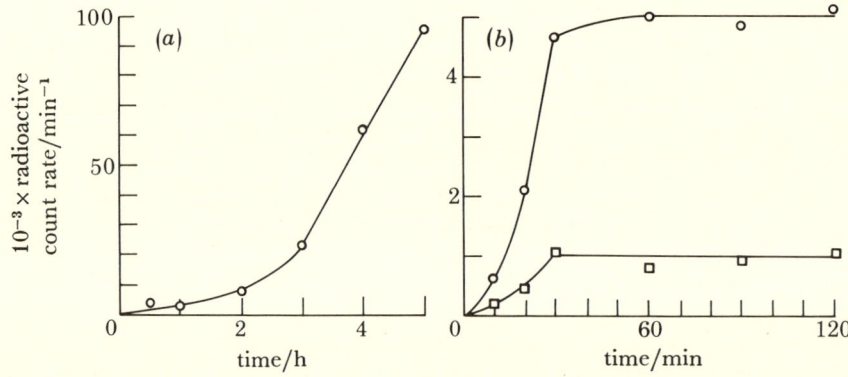

FIGURE 3. Time course of incorporation of L-[1-^{3}H]fucose into polysaccharide (a) (density 1.63 g cm^{-3}) and glycoprotein fractions (b) (densities 1.37 (○) and 1.55 (□) g cm^{-3}) of the maize root tip (Green & Northcote 1978).

their radioactive composition determined (table 4). It can be seen that the incorporation of radioactivity into the sugars of the glycoproteins resembles that of the slime polysaccharides. The polymers that were analysed were those that were isolated directly from the root tip. The glycoproteins were present in the material derived from the membrane fractions only.

The time courses for the incorporation of L-[1-^{3}H]fucose into polysaccharide and glycoprotein fractions are shown in figure 3. If the roots were incubated for 10 min in radioactive fucose and then left in a non-radioactive fucose solution for up to 150 min, the radioactivity was chased out of the glycoprotein and accumulated in the free polysaccharide (figure 4). The glycoproteins could therefore be intermediates and carriers of the carbohydrate that finally appeared as the free slime polysaccharides excreted from the root tip.

It was possible that the two glycoproteins were structural polymers of the membranes rather than intermediates involved in the synthesis and transport of polysaccharide. However, the great size of the saccharide portion of the glycoproteins (larger than 10 kDa, a d.p. of at least 60 sugar residues) and the very low content of mannose together with the similarity of the sugar composition of the glycoprotein and polysaccharide, make this unlikely.

One of the glycoproteins (density 1.3 g cm^{-3}) we have investigated in more detail to determine the linkage between the carbohydrate and protein. The glycoprotein can be β-eliminated if it is

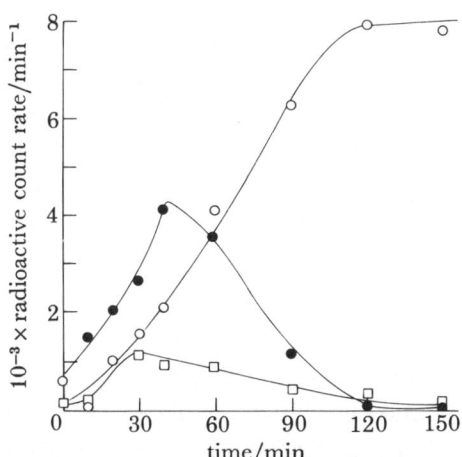

FIGURE 4. Incorporation of L-[1-^{3}H]fucose into polysaccharide (density 1.63 g cm^{-3} ((o) and glycoprotein (density 1.37 (●) and 1.55 (□) g cm^{-3}) fractions of the maize root tip during a pulse–chase experiment (Green & Northcote 1978).

treated with 0.5 M sodium hydroxide at 20 °C for 24 h so that the protein and oligosaccharide are separated. The material after β-elimination could be reduced with sodium[^{3}H] borohydride and, after hydrolysis of the radioactively labelled product, radioactive α-aminobutyric acid was detected. Similarly when the glycoprotein was made radioactive by incubating roots in L-[U-^{14}C]threonine and the glycoprotein was β-eliminated and hydrolysed, radioactive α-amino-butyric acid was again detected in the hydrolysate. The sugar at the attachment point between the protein and oligosaccharide was detected by carrying out the β-elimination in the presence of [^{3}H]sodium borohydride and hydrolysis of the products. Radioactive xylitol was the only sugar alcohol that was found. Thus a xylose–threonine linkage was present in this glycoprotein; no evidence for an attachment to the hydroxy group of serine was found (Green & Northcote 1978). The other glycoprotein, at a density of 1.55 g cm^{-3}, was found not to be susceptible to β-elimination.

GLYCOLIPIDS PRESENT IN THE MEMBRANE SYSTEM

Two types of experiments were carried out. Either maize roots were incubated in L-[1-^{3}H]-fucose and the membranes were then isolated and analysed, or membranes were prepared from root tips and these were incubated with GDP-L-[U-^{14}C]fucose or UDP-D-[U-^{14}C]glucose (Green & Northcote 1979a). The radioactively labelled lipids that were synthesized in vivo and in vitro were extracted and separated into polar and neutral component. It was significant that no lipids containing fucose were found in parts of the root away from the root-tip in the experiments in vivo, or in membranes obtained from higher up the root in the experiments in vitro.

The polar lipids had the characteristics of polyprenyl diphosphate fucose, and glucose derivatives and possibly some polyprenyl monophosphate glucose derivatives were present. The neutral lipids were fucose or glucose sterol glycosides; the sterol fucosides were again only found in the preparations from the root tip.

Membranes were fractionated into those sedimenting at 20 000 g and those at 100 000 g, and it was shown by enzyme characterization that the 100 000 g pellet concentrated the endoplasmic reticulum and the 20 000 g pellet concentrated the membranes of the Golgi apparatus. Most of the transglycosylases that gave rise to the polar lipids for both glucose and fucose were located in the 100 000 g pellet. The sterol fucosides were mainly synthesized by the 20 000 g pellet and the activity was probably associated with the Golgi apparatus, whereas the sterol glucosides were synthesized by both membrane fractions.

A more detailed study of the polar lipids derived by labelling *in vivo* and extraction of the dried membranes with mixtures of chloroform and methanol was made. The polar lipids were chromatographed on a column of DEAE-cellulose. The fractions from the column were hydrolysed by 0.1 M hydrochloric acid in tetrahydrofuran (80 % by volume) and the oligosaccharides analysed on a Biogel P-2 column to determine their size. A complete acid hydrolysis to determine their composition was also carried out. Some of the oligosaccharides were built up of at least nine sugar units. The composition of the unfractionated oligosaccharides resembled that of the slime polysaccharide and those of the carbohydrates of the glycoproteins. Some of the oligosaccharide contained at least four different sugars including galacturonic acid, glucose, arabinose and fucose; however, some did not contain fucose. The oligosaccharides were present as polyprenyl diphosphate derivatives (Dixon & Northcote 1983).

TRANSGLYCOSYLASES

There are at least two types of fucosyl transglycosylase found on the membrane system of the cells of the root tip. One transfers fucose to polyprenyl phosphate and the other transfers fucose to polysaccharide or glycoprotein. These were investigated by experiments *in vivo* in which L-[1-^{3}H]fucose was incubated with intact roots and the membranes were isolated and characterized, and fucosyl derivatives extracted from them. They were also studied by isolation of the membranes from the cells and incubation of the membranes with GDP-L-[U-^{14}C]fucose. The membranes were carefully isolated and characterized from a discontinuous sucrose gradient into fractions at the interfaces of 8–18 % sucrose, 18–25 %, 25–29 %, 29–33 %, 33–39 %, 39–45 %, greater than 45 %, and also a supernatant fraction. Membranes were prepared in the presence of Mg^{2+}, which maintained the attachment of ribosomes to the endoplasmic reticulum, and also in the presence of EDTA, which removed most of the ribosome complex. In this way the membranes of the endoplasmic reticulum were displaced in the density gradient and could be unambiguously identified (Green & Northcote 1979b).

The GDP fucose:polyprenylphosphate transfucosylase occurred in the endoplasmic reticulum, and the fucosyl transferase that transfers fucose to polymer occurred mainly in the Golgi apparatus, but there was a significant amount of activity associated with the endoplasmic reticulum. Whether the transfucosylase that transfers fucose to the polymers uses a lipid intermediate as a substrate at either location is not known.

Membrane fusion and secretion by the membrane system

To investigate the possible mechanisms of membrane fusion an *in vitro* system was devised in which the membranes from the root tip were isolated into four membrane fractions and soluble material on a discontinuous sucrose gradient. The membrane fractions enriched in endoplasmic reticulum (14–25 % sucrose interface), Golgi apparatus (25–34 %), plasma membrane (34–39 %) and mitochondria (39–45 %) were isolated and characterized by their enzymic properties

Table 5. Transfer of radioactivity between membrane fractions one of which was non-radioactive and the other made radioactive with D-[U-^{14}C]glucose

(Roots were incubated with D-[U-^{14}C]glucose *in vivo*. Membrane fractions were prepared from radioactive roots (*) and from non-radioactive roots (^{0}). These were mixed at 30 °C for 40 min in the presence of 1 mM Ca^{2+} and then refractionated by centrifugation. The distribution of radioactivity in the fractions was determined. Recovery was calculated by comparing total radioactive count rate with the initial count rate obtained by pelleting a radioactive fraction.)

fraction	radioactivity (%) from refractionated membranes after mixing			
	G.a.* with p.m.^{0}	G.a.* alone	G.a.^{0} with p.m.*	p.m.* alone
Golgi apparatus-rich (G.a.)	47.9	89.1	32.8	4.0
plasma membrane-rich (p.m.)	49.4	8.3	47.8	75.1
mitochondria-rich	0.9	0.9	10.7	12.0
remainder fraction	1.8	1.7	8.7	8.9
total count rate obtained from all the fractions/min^{-1}	14950	13650	7180	6350
recovery (%)	90.0	83.0	94.8	83.8

and ultrastructural appearance (Baydoun & Northcote 1980a). Maize root tips were incubated with radioactive glucose or choline and membranes were prepared from radioactive and non-radioactive roots. The membranes were then mixed, for example radioactive Golgi apparatus with the non-radioactive plasma membrane fraction at 30 °C for 40 min, and then refractionated. Radioactivity was transferred between the membrane fractions with all the radioactive markers used. Control experiments without mixing showed that the membranes could be recovered at the appropriate densities. The transfer of radioactivity between mixed membrane fractions has enabled a quantitative system to be developed to study membrane fusion *in vitro* between a Golgi apparatus-rich fraction and one enriched in plasma membranes (tables 5 and 6) (Baydoun & Northcote 1980b).

Membrane fusion was found to be dependent on time, temperature, Mn^{2+} and Ca^{2+}. Mn^{2+} was as effective as Ca^{2+} in promoting membrane fusion, but other divalent cations including Mg^{2+} had a moderate effect, or none.

Trypsin treatment of mixed membrane fractions before the addition of Ca^{2+} inhibited their ability to fuse. It also resulted in a selective and progressive elimination of a characteristic intense polypeptide band seen when the membrane proteins were displayed by electrophoresis on a sodium dodecyl sulphate (SDS) polyacrylamide gel. Thus, the fusion of the membranes that was enhanced by the presence of Ca^{2+} was probably dependent on the presence of this particular membrane protein. It was not removed by chymotrypsin or thermolysin. It appeared to be an integral membrane protein with an exposed portion of the peptide chain to the outside of the membrane, and it was this that was degraded by the trypsin.

The results indicated that Ca^{2+} could probably form bridges between mixed membrane vesicles through negative charges on surface proteins or glycoproteins, or that there were present specific polypeptides that bind Ca^{2+} ions. The bridges would bring membranes into close contact, a prerequisite for membrane fusion. Trypsin treatment did not completely abolish membrane fusion in the presence of Ca^{2+}. This could be due to the presence of other proteins that were not removed by trypsin or that Ca^{2+} could bind to acidic phospholipids to bring membranes in contact.

TABLE 6. TRANSFER OF RADIOACTIVITY BETWEEN MEMBRANE FRACTIONS ONE OF WHICH WAS NON-RADIOACTIVE AND THE OTHER MADE RADIOACTIVE WITH [Me-^{14}C]CHOLINE CHLORIDE

(Roots were incubated with [Me-^{14}C]choline chloride *in vivo*. Membrane fractions were prepared from radioactive roots (*) and from non-radioactive roots (⁰). These were mixed at 30 °C for 40 min in the presence of 1 mM Ca^{2+} and then refractionated by centrifugation. The distribution of radioactivity in the fractions was determined.)

fraction	radioactivity (%) from refractionated membranes after mixing			
	G.a.* with p.m.⁰	G.a.* alone	G.a.⁰ with p.m.*	p.m.* alone
Golgi apparatus-rich	48.1	83.0	32.8	6.7
plasma membrane-rich	46.3	10.6	53.3	83.1
remainder fraction	5.6	6.4	13.9	10.2
total count rate obtained from all the fractions/min^{-1}	18880	18700	8520	8150
recovery (%)	99.0	98.1	97.8	93.5

Sodium deoxycholate was used to solubilize the proteins of mixed membrane fractions. The extracted proteins were analysed by non-SDS polyacrylamide gel electrophoresis: at least four bands were formed that could be cut from the gel. One of these bands had the same mobility as the polypeptide removed by trypsin when it was re-run electrophoretically in a gel with SDS. The molecular mass of this band was 36 kDa. The non-SDS polyacrylamide gels were stained for various enzymic activities, and the position at which the trypsin-sensitive polypeptide band ran showed that there was a protein present that was a Ca^{2+} and Mg^{2+} activated ATPase (Baydoun & Northcote 1981).

SUMMARY OF THE MECHANISM OF SYNTHESIS OF SLIME AND ITS POSSIBLE CONTROL

The slime secreted from the roots of maize is a complex of at least three polysaccharides. It is assembled and secreted by the membrane system of the outer root cap cells. One of the polysaccharides is acidic, containing galacturonic acid and also large quantities of fucose. Glycoproteins and polyprenyl diphosphate oligosaccharides of similar sugar composition to this polysaccharide have been identified in the membrane system of the cells. In addition, radioactive pulse–chase experiments have indicated that the glycoproteins could be intermediates during the assembly and synthesis of the polysaccharides.

It is therefore possible that the polysaccharide is assembled by the transfer of a regular sequence of sugars in the form of an oligosaccharide of at least nine sugar residues synthesized on to a polyprenyl acceptor. This type of synthesis has recently been shown to occur during the biosynthesis of an acidic polysaccharide secreted by *Acetobacter xylinum* (Couso *et al.* 1982) and also during the formation of xanthan gum secreted by *Xanthomonas campestris* (Ielpi *et al.* 1981 *a*, *b*).

In the maize root tip the synthesis of the lipid-oligosaccharide probably starts in the endoplasmic reticulum, and transfer onto glycoprotein occurs in the Golgi apparatus. However, the early stages of this polymerization process, such as the initial transfer of an oligosaccharide or sugar to the protein to give the nascent glycoprotein, probably occur at the endoplasmic reticulum.

Work that we have carried out with suspension cultures of sycamore cells show that these cells secrete polysaccharides into the growth medium. These polysaccharides are similar to those present in the matrix material of the wall and synthesized and secreted into the wall by exopinocytotic vesicles derived from the Golgi apparatus. With this system it can be shown that the application of Ca^{2+} immediately stimulates the secretion within 10 s from the addition of the divalent ion (Morris & Northcote 1977). Thus for secretion at the plasma membrane there is an immediate control, and a rate-limiting factor in secretion is the fusion of the vesicles at the cell surface. Such a control at the cell surface to increase or decrease the rate of secretion of the polysaccharides must be correlated with their rate of synthesis within the membrane systems. We have shown that the rate of synthesis of a number of polysaccharides produced by plant cells and secreted into the wall is controlled at the synthetase steps, and during differentiation of the cell these synthetases are induced or repressed at the various developmental stages. With the polymers secreted in the slime of maize root tips and maybe in the synthesis of other polysaccharides where lipid or protein acceptors could be involved and on which the polymers are in part assembled, control could be exerted at the several transglycosylates involved at each stage (Northcote 1982 b).

REFERENCES

Allen, D. M. & Northcote, D. H. 1975 The scales of *Chrysochromulina chiton*. *Protoplasma* **83**, 389–412.

Barlow, P. W. 1975 The root cap. In *The development and function of roots* (ed. J. G. Torrey & D. T. Clarkson), pp. 21–54. London: Academic Press.

Baydoun, E. A. H. & Northcote, D. H. 1980a Isolation and characterization of membranes from the cells of maize root-tips. *J. Cell Sci.* **45**, 147–167.

Baydoun, E. A. H. & Northcote, D. H. 1980b Measurement and characteristics of fusion of isolated membrane fractions from maize root-tips. *J. Cell Sci.* **45**, 169–186.

Baydoun, E. A. H. & Northcote, D. H. 1981 The extraction from maize (*Zea mays*) root cells of membrane-bound protein with Ca^{2+}-dependent ATPase activity and its possible role in membrane fusion *in vitro*. *Biochem. J.* **193**, 781–792.

Bowles, D. J. & Northcote, D. H. 1972 The sites of synthesis and transport of extracellular polysaccharides in the root tissue of maize. *Biochem. J.* **130**, 1133–1145.

Bowles, D. J. & Northcote, D. H. 1974 The amounts and rates of export of polysaccharides found within the membrane system of maize root cells. *Biochem. J.* **142**, 139–144.

Bowles, D. J. & Northcote, D. H. 1976 The size and distribution of polysaccharides during their synthesis within the membrane system of maize root cells. *Planta* **128**, 101–106.

Couso, R. O., Ielpi, L., Garcia, R. C. & Dankert, M. A. 1982 Biosynthesis of polysaccharides in *Acetobacter xylinum* sequential synthesis of a heptasaccharide diphosphate prenol. *Eur. J. Biochem.* **123**, 617–627.

Dixon, W. T. & Northcote, D. H. 1983 Glycolipids and glycoproteins during maize-root slime synthesis. In preparation.

Grant, G. T., McNab, C., Rees, D. A. & Skerrett, R. J. 1969 Seed mucilages as examples of polysaccharide denaturation. *Chem. Commun.*, pp. 805–806.

Green, J. R. & Northcote, D. H. 1978 The structure and function of glycoproteins synthesised during slime-polysaccharide production by membranes of the root-cap cells of maize (*Zea mays*). *Biochem. J.* **170**, 599–608.

Green, J. R. & Northcote, D. H. 1979a Polyprenyl phosphate sugars synthesised during slime-polysaccharide production by membranes of the root-cap cells of maize (*Zea mays*). *Biochem. J.* **178**, 661–671.

Green, J. R. & Northcote, D. H. 1979b Location of fucosyl transferases in the membrane system of maize root cells. *J. Cell Sci.* **40**, 235–244.

Harris, P. J. & Northcote, D. H. 1971 Polysaccharide formation in plant Golgi bodies. *Biochim. biophys. Acta* **237**, 56–64.

Ielpi, L., Couso, R. & Dankert, M. 1981a Lipid-linked intermediates in the biosynthesis of xanthan gum. *FEBS Lett.* **130**, 253–256.

Ielpi, L., Couso, R. & Dankert, A. 1981b Xanthan gum biosynthesis. *Biochem. biophys. Res. Commun.* **102**, 1400–1408.

Kirby, K. G. & Roberts, R. M. 1971 The localized incorporation of ^{3}H-L-fucose into cell wall polysaccharides of the cap and epidermis of corn roots. Autoradiographic and biosynthetic studies. *Planta* **99**, 211–221.

Morris, M. R. & Northcote, D. H. 1977 Influence of cations at the plasma membrane in controlling polysaccharide secretion from sycamore suspension cells. *Biochem. J.* **166**, 603–618.

Murashige, T. & Skoog, F. 1962 A revised medium for rapid growth and bioassays with tobacco tissue cultures. *Physiol. Pl.* **15**, 473–496.

Northcote, D. H. 1982a Macromolecular aspects of cell wall differentiation. In *Encyclopedia of plant physiology* (new series), vol. 14A (ed. D. Boulter), pp. 638–655. Berlin: Springer-Verlag.

Northcote, D. H. 1982b Control of enzyme activity during plant cell development. In *Differentiation* in vitro. (*Br. Soc. Cell Biol. Symp.* vol. 4) (ed. M. M. Yeoman & D. E. S. Truman), pp. 49–64. Cambridge University Press.

Northcote, D. H. & Pickett-Heaps, J. D. 1966 A function of the Golgi apparatus in polysaccharide synthesis and transport in the root-cap cells. *Biochem. J.* **98**, 159–167.

Paull, R. E. & Jones, R. L. 1975 Studies on the secretion of maize root-cap slime. II. Localization of slime production. *Pl. Physiol.* **56**, 307–312.

Paull, R. E. & Jones, R. L. 1976 Studies on the secretion of maize root-cap slime. IV. Evidence for the involvement of dictyosomes. *Pl. Physiol.* **57**, 249–256.

Rougier, M. 1976 Sécrétion de polysaccharides dans l'apex radiculaire de mais: étude radioautographique par incorporation de fucose tritié. *J. Microsc., Paris* **26**, 161–166.

Rougier, M. 1981 Secretionary activity of the root-cap. In *Encyclopedia of plant physiology* (new series), vol. 13B (*Plant carbohydrates II*) (ed. W. Tanner & F. A. Loewus), pp. 542–574. Berlin: Springer-Verlag.

Wright, K. & Northcote, D. H. 1974 The relationship of root-cap slimes to pectins. *Biochem. J.* **139**, 525–534.

Wright, K. & Northcote, D. H. 1975 An acidic oligosaccharide from maize slime. *Phytochemistry* **14**, 1793–1798.

Wright, K. & Northcote, D. H. 1976 Identification of β1→4 glucan chains as part of a fraction of slime synthesized within the dictyosomes of maize root caps. *Protoplasma* **88**, 225–239.

Wright, K., Northcote, D. H. & Davey, R. H. 1976 Preparation of rat epididymal α-L-fucosidase free from other glycosidases: its action on root-cap slime from *Zea mays* L. *Carbohydr. Res.* **47**, 141–150.

Phil. Trans. R. Soc. Lond. B **300**, 207–223 (1982)

Printed in Great Britain

Genetic and biochemical studies of asparagine-linked oligosaccharide assembly

By M. D. Snider, T. C. Huffaker, J. R. Couto and P. W. Robbins

Center for Cancer Research and Department of Biology, Massachusetts Institute of Technology,
Cambridge, Massachusetts 02139, *U.S.A.*

The formation of *N*-glycosidic linkages of eukaryotic glycoproteins involves the assembly of a specific lipid-linked precursor oligosaccharide in the endoplasmic reticulum. This oligosaccharide is transferred from the lipid carrier to appropriate asparagine residues during protein synthesis. The protein-linked oligosaccharide then undergoes processing reactions that include both removal and addition of carbohydrate residues.

In this paper we report recent studies from our laboratory on the synthesis of asparagine-linked oligosaccharides. In the first part we describe the isolation and characterization of temperature-sensitive mutants of yeast blocked at specific stages in the assembly of the lipid-linked oligosaccharide. In addition, we are using these mutants to clone the genes for the enzymes in this pathway by complementation of the temperature-sensitive phenotype. The second part deals with the topography of asparagine-linked oligosaccharide assembly. Our studies on the transmembrane movement of sugar residues during the assembly of secreted glycoproteins from cytoplasmic precursors are presented. Finally, experiments on the control of protein-linked oligosaccharide processing are described. Recent data are presented on the problem of how specific oligosaccharides are assembled from the common precursors at individual sites on glycoproteins.

Introduction

The asparagine-linked oligosaccharides of eukaryotic glycoproteins are assembled in two distinct stages (reviewed in Hubbard & Ivatt 1981; Struck & Lennarz 1980). The first step involves the assembly of a large lipid-linked precursor oligosaccharide containing *N*-acetylglucosamine (GlcNAc), mannose (Man) and glucose (Glc) (figure 1). This species then acts as the donor in the transfer of the oligosaccharide from the lipid carrier to nascent polypeptide chains. In the second stage, the initial common protein-linked oligosaccharide is extensively modified to yield the diverse array of final products.

The precursor oligosaccharide-lipid is assembled in the membrane of the endoplasmic reticulum (e.r.) from the sugar nucleotides UDP-GlcNAc, GDP-Man, and UDP-Glc (figure 2). The first sugar is added as GlcNAc-P, to the polyisoprenoid lipid, dolichyl phosphate (Dol-P), with the formation of the pyrophosphate lipid–sugar linkage. Then six more sugars are added direct from the sugar nucleotides to yield the key intermediate $Man_5GlcNAc_2$-PP-Dol. The seven outer sugars are added via the phosphate-linked intermediates, Dol-P-Man and Dol-P-Glc. These two compounds are made from Dol-P and the respective sugar nucleotides (figure 2). Upon completion, the $Glc_3Man_9GlcNAc_2$ oligosaccharide can be transferred to growing polypeptide chains in the membrane of the e.r.

The second stage of asparagine-linked oligosaccharide synthesis begins with the removal of the three glucose residues from the newly made glycopeptide in the e.r. Then, in the Golgi apparatus, Man residues are removed, and terminal sugars, phosphate and sulphate residues are added to yield mature structures. While the reactions of the first stage of the glycosylation pathway appear to be shared by nearly all eukaryotes, the reactions in the second stage are specific for individual species, cell types and glycoproteins. These reactions are responsible for the synthesis of the diverse array of final products from a single common precursor.

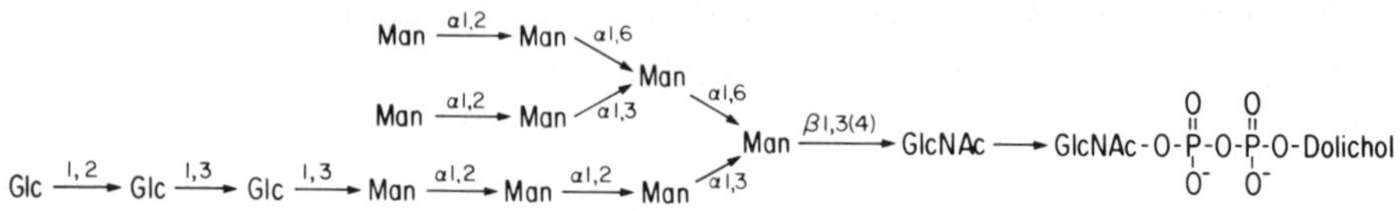

FIGURE 1. Structure of the dolichol-linked precursor oligosaccharide-lipid.

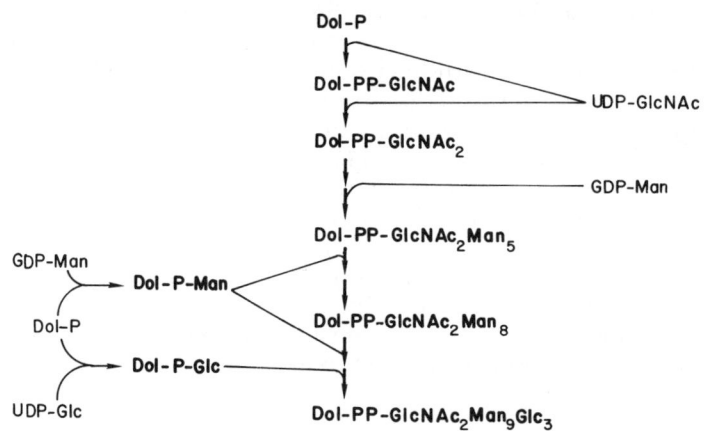

FIGURE 2. Pathway of oligosaccharide-lipid synthesis.

In this paper, we report recent studies from our laboratory on the synthesis of asparagine-linked oligosaccharides. The first section deals with the isolation of mutants in yeast that are defective in lipid-linked oligosaccharide assembly and the use of these mutants to clone the affected genes. The second section examines the topography of asparagine-linked oligosaccharide assembly and the transmembrane movement of sugar residues during glycoprotein synthesis. Finally, experiments on the control of the second stage of glycosylation are described.

YEAST MUTANTS DEFICIENT IN PROTEIN GLYCOSYLATION

We believe that the characterization of mutants will be useful in unravelling the pathway of asparagine-linked glycosylation and have been working with the yeast *Saccharomyces cerevisiae*. Although recent studies with mammalian cell mutants have made valuable contributions to the understanding of protein glycosylation (Chapman *et al.* 1979; Hunt 1980), the relative ease of obtaining mutants in yeast has allowed us to use a genetic and biochemical approach that has been less feasible in animal cell systems. Like higher eukaryotic cells, yeast cells make lipid-linked $Glc_3Man_9GlcNAc_2$ (Trimble *et al.* 1980; Lehle 1980), which is transferred *en bloc* to protein and then modified by excision of the glucose residues (Parodi 1979; Lehle 1980).

Whereas further processing in animal cells can involve the removal of up to six mannose residues and the addition of sugars such as *N*-acetylglucosamine, galactose, fucose and sialic acid, yeast cells add mannose residues and produce oligosaccharides ranging in size from 13 mannose residues, in carboxypeptidase Y (Trimble & Maley 1977; Hasilik & Tanner 1978), to over 50, in invertase (Tarentino *et al.* 1974; Lehle *et al.* 1979). The addition of these outer mannose residues has been studied extensively by Ballou and coworkers, and mutants that fail to synthesize the outer chain, the *mnn* mutants, have been isolated (Ballou 1980).

We have been interested in obtaining mutants affected in the early steps of the glycosylation pathway, which yeast cells have in common with higher eukaryotic cells. These include mutants blocked in the synthesis of the lipid-linked oligosaccharide precursor, its co-translational transfer to protein, and the first steps of post-translational processing. A commonly used method for obtaining protein glycosylation mutants in yeast and animal cells is the selection of cells unable to bind either lectins or antibodies that recognize cell–surface components (Stanley 1980; Ballou 1980). Although this approach has produced many mutants with defects in the late stages of oligosaccharide processing, only a few have been obtained that are affected in the earlier steps of protein glycosylation. This suggests that many of these early steps may be necessary for cell viability. We have therefore developed procedures for isolating mutants in yeast that contain temperature-sensitive defects in asparagine-linked glycosylation. A [³H]mannose suicide selection was used to enrich for these mutants. The surviving cells were screened at the non-permissive temperature for a decreased ability to incorporate [³H]mannose and for defects in glycosylation of the secreted protein invertase. These methods have allowed us to identify a number of mutants that are blocked at various stages in the assembly of the lipid-linked oligosaccharide precursor. In addition we are cloning the genes for the enzymes in this pathway by complementation of the temperature-sensitive mutant phenotype.

[³H]*Mannose suicide selection*

To obtain mutants in yeast that are temperature-sensitive for the ability to glycosylate proteins, we employed a [³H]mannose suicide selection. It was predicted that cells blocked in protein glycosylation would incorporate less mannose than wild-type cells and would therefore be less likely to sustain lethal radiation damage. To ensure that the selection was specific for mannose incorporation, mannose tritiated in the 2-position was used. Catabolism of [2-³H]-mannose results in removal of the label as tritiated water when mannose phosphate isomerase converts mannose 6-phosphate to fructose 6-phosphate.

The suicide selection involved labelling ethyl methanesulphonate mutagenized yeast cells with [³H]mannose for 30 min at the non-permissive temperature. Washed cells were frozen in 25 % glycerol and stored at − 80 °C to allow accumulation of radiation damage. Periodically, aliquots were thawed and analysed for cell survival. Figure 3 shows the time course of cell killing for a culture labelled with a radioactive count of approximately 50 min⁻¹ per cell of [³H]mannose. During the first 27 days, the rate of cell killing followed first-order kinetics. After 35 days, the survival was about 0.2 %.

Screening for mutants

To screen for mutants blocked in protein glycosylation, colony fluorography was used to identify colonies that failed to incorporate wild-type levels of [³H]mannose after 30 min at the non-permissive temperature. Since any temperature-sensitive defect that lowers the rate

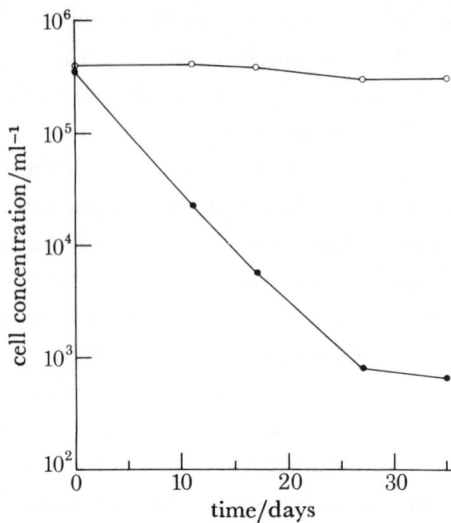

FIGURE 3. Cell survival during [³H]mannose suicide selection. Mutagenized yeast cells were labelled with [³H]-mannose at the non-permissive temperature and stored at − 80 °C (Huffaker & Robbins 1982). The number of viable cells per millilitre were determined during the course of the suicide by scoring for growth on plates at the permissive temperature. ●, [³H]Mannose-labelled cells; ○, unlabelled cells included as a control.

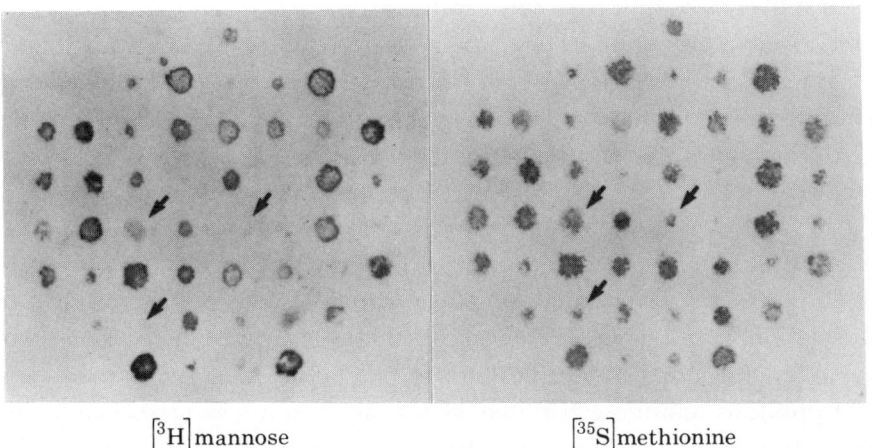

$\left[^3\mathrm{H}\right]$mannose $\qquad$ $\left[^{35}\mathrm{S}\right]$methionine

FIGURE 4. Colony screen for incorporation of [³H]mannose and [³⁵S]methionine. Colonies were replica-plated onto two filter papers and assayed for their ability to incorporate [³H]mannose and [³⁵S]methionine at the non-permissive temperature (Huffaker & Robbins 1982). Fluorograms of [³H]mannose and [³⁵S]methionine-labelled replicas from a single plate are shown. Colonies that incorporated [³⁵S]methionine but failed to incorporate [³H]mannose are indicated by arrows.

of protein synthesis could indirectly affect the rate of protein glycosylation, colonies were also screened for their ability to incorporate [³⁵S]methionine. Cells that survived 35 days of the [³H]mannose suicide were grown on plates and the colonies were replica-plated onto two filter papers. One filter was incubated with [³H]mannose and the other with [³⁵S]methionine in medium at the non-permissive temperature. After fixing the cells with glutaraldehyde, the filters were rinsed and impregnated with 2,5-diphenyloxazole. Fluorography of the filters was used to determine the amount of incorporation of the labelled compounds (figure 4). Approximately 5–10 % of the colonies screened incorporated substantially decreased levels of [³H]-

mannose, while [³⁵S]methionine incorporation was nearly normal. These were chosen as the most likely to be mutants with glycosylation defects.

Cells that incorporated low amounts of [³H]mannose in the colony screen were next assayed for their ability to glycosylate the secreted form of invertase. The synthesis of secreted invertase is repressed when cells are grown in the presence of 2 % glucose, but is derepressed when cells are grown in low-glucose medium (Perlman & Halvorson 1981). Cultures of mutants identified by the colony screen were derepressed for invertase production at the non-permissive temperature. Cell extracts were prepared and run on polyacrylamide gels. After the gels had been stained for invertase activity, the extent of invertase glycosylation was estimated by observing its electrophoretic mobility (figure 5). Wild-type cells (lane o) synthesized secreted invertase, which migrated as a broad band at the top of the gel. The non-glycosylated cytoplasmic form

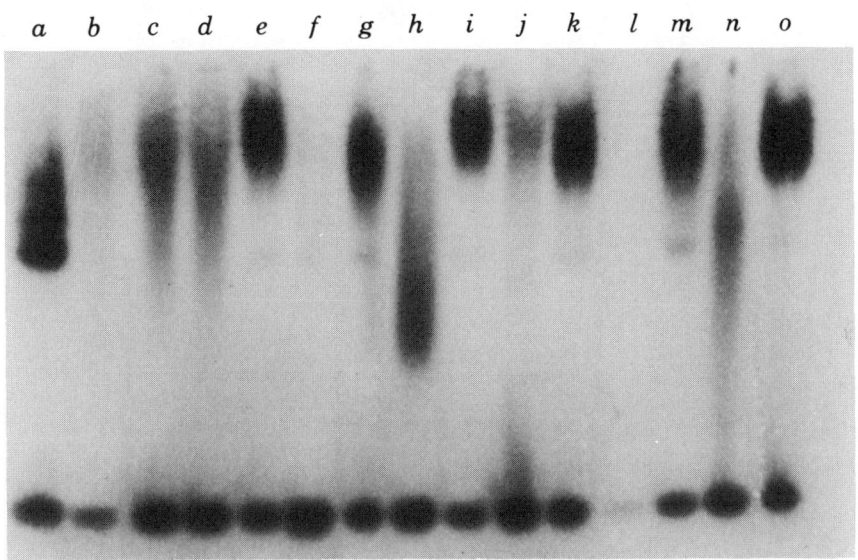

FIGURE 5. Assay for glycosylation of invertase. Cultures were derepressed for invertase synthesis at the non-permissive temperature. Cell extracts were run on a polyacrylamide gel and the gel was stained for invertase activity (Huffaker & Robbins 1982). Lane o, wild type; lanes a–n, mutants identified by the colony screen.

of invertase ran near the bottom (Meyer & Matile 1974). The mutants can be divided into three classes. Some produced normal amounts of fully glycosylated invertase (lanes e, g, i, k and m). However, a few synthesized much less of the fully glycosylated invertase than wild-type cells (lanes b, f, j and l), and others appeared to make incompletely glycosylated invertase based on the altered mobilities observed (lanes a, c, d, h and n).

Mutants blocked in the synthesis of the precursor lipid-linked oligosaccharide

Mutants that failed to glycosylate invertase normally are likely to contain general protein glycosylation defects. To determine whether any of these mutants are affected in the synthesis of the precursor lipid-linked oligosaccharide, cultures were labelled with [³H]mannose at the non-permissive temperature for 10 min. The lipid-linked oligosaccharides were extracted and subjected to mild acid hydrolysis. The released oligosaccharides were then analysed by gel filtration chromatography on Bio-Gel P-4 (minus 400 mesh). In wild-type cells Glc_3Man_9-$GlcNAc_2$ was the major lipid-linked oligosaccharide labelled, with smaller amounts of $Man_8GlcNAc_2$, $Man_7GlcNAc_2$ and $Man_5GlcNAc_2$ (figure 6a). Of the 50 mutants examined, 22 failed to synthesize detectable amounts of $Glc_3Man_9GlcNAc_2$ or accumulated other intermediates in the synthesis of the lipid-linked precursor, or both. Figure 6 shows gel filtration profiles of representative mutants that accumulate as major peaks $Man_9GlcNAc_2$, $Man_8GlcNAc_2$,

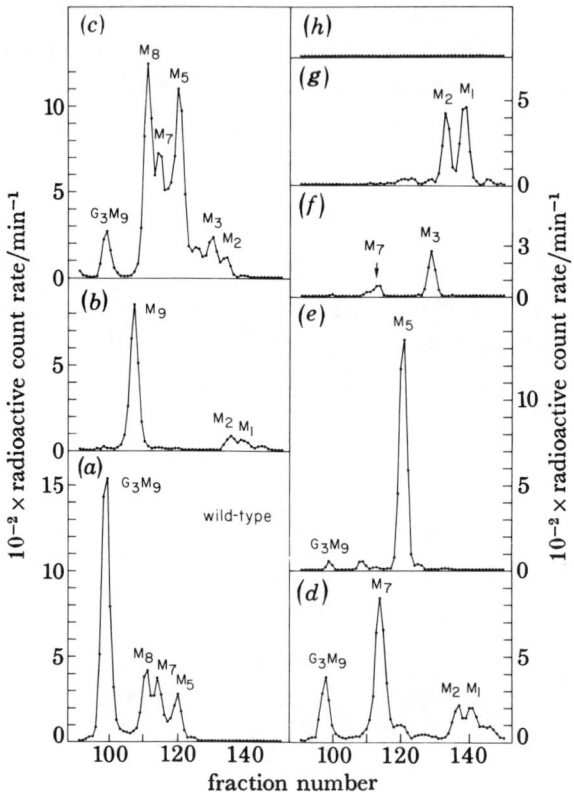

FIGURE 6. Gel filtration chromatography of lipid-linked oligosaccharides labelled *in vivo*. Wild-type (a) and mutant cells (b–h) were labelled with [³H]mannose at the non-permissive temperature and the lipid-linked oligosaccharides were extracted, hydrolysed and analysed by gel filtration on Bio-Gel P4 (minus 400 mesh) (Huffaker & Robbins 1982). G, Glc; M, Man. All species contain two GlcNAc residues at the reducing end.

$Man_7GlcNAc_2$, $Man_5GlcNAc_2$, $Man_3GlcNAc_2$ or $Man_2GlcNAc_2$. One mutant fails to incorporate any detectable amount of [³H]mannose into lipid-linked oligosaccharides (figure 6h). This mutant (*alg1-1*) has been shown to be blocked in the addition of the first mannose residue to the oligosaccharide-lipid (Huffaker & Robbins 1982).

Complementation analysis can be readily performed in yeast by mating two haploid mutant cells and examining the phenotype of the resultant diploid. For those mutations that are lethal at the non-permissive temperature, complementation analysis has been done by assaying diploids for temperature-sensitive growth. We have found one complementation group that contains nine members and includes mutants that accumulate predominantly $Man_8GlcNAc_2$ and the one mutant that accumulates $Man_7GlcNAc_2$. A second complementation group contains three members, each of which accumulates $Man_2GlcNAc_2$ and $Man_1GlcNAc_2$. For those mutations that are not temperature-sensitive lethal, complementation analysis is being performed by examining the lipid-linked oligosaccharides synthesized by the relevant diploids at the non-permissive temperature.

The [³H]mannose suicide and the screens described have proved to be useful in isolating mutants blocked in lipid-linked oligosaccharide assembly, and it seems reasonable to expect that this pathway can be saturated with mutations. Approximately half of the mutants that fail to glycosylate invertase normally synthesize the complete lipid-linked oligosaccharide $Glc_3Man_9GlcNAc_2$. Some of these mutants may contain a defective oligosaccharide transferase, the enzyme that transfers the precursor oligosaccharide from lipid to protein. Alternatively, these mutants may be blocked in processing the protein-linked oligosaccharide, preventing its elongation by the enzymes that add the mannose residues of the outer chain. While we have not yet identified mutants of this type, procedures similar to those used here should allow their isolation in the future.

Cloning of genes involved in oligosaccharide synthesis

We are currently attempting to isolate the genes in the asparagine-linked glycosylation pathway. Our experimental approach is as follows. (1) Isolation and characterization of temperature-sensitive lethal mutations in different steps of the pathway. This has been described in detail

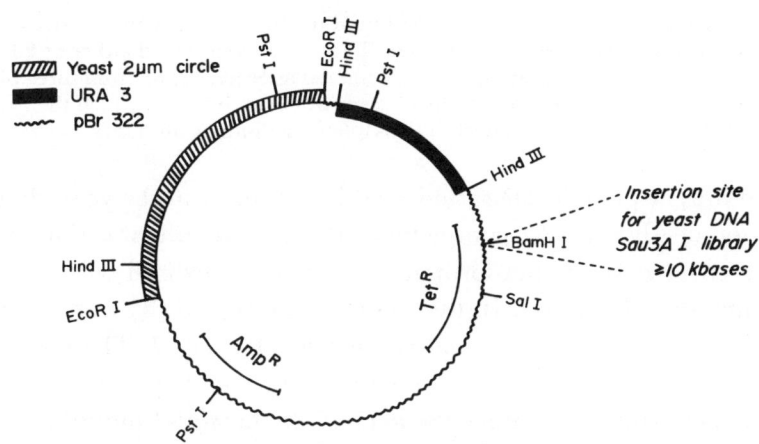

FIGURE 7. YEp24 (Botstein *et al.* 1979) transforms both yeast and *E. coli.* It contains part of the 2 μ yeast plasmid with its origin of replication, the yeast URA3 gene, and pBR322. It complements both the *ura3* mutation in yeast and the *pyr* mutation in *E. coli*, and contains the *Amp*[r] and *Tet*[r] markers for drug selection in *E. coli*. It has two unique restriction sites, *Bam*HI and *Sal*I. The *Bam*HI site was used for the insertion of *Sau*3AI partial restriction digests for the construction of the yeast library (Carlson & Botstein 1982).

above. (2) Transformation of these mutants with libraries (Maniatis *et al.* 1978) of wild-type yeast DNA. A library is a collection of plasmids each containing a different fragment of DNA randomly excised from a whole genome. Thus a library should in principle contain all genes of a given strain represented. (3) Selection of transformants in which the temperature-sensitivity has been complemented by a fragment of the DNA library. It is then reasonable to conclude that such transformants acquired a plasmid containing the gene in question, and that its expression is now enabling the transformants to grow at the non-permissive temperature. (4) Subcloning and characterization of the gene. Using this approach we have isolated a plasmid, pJC1, which complements the *alg1-1* mutation (Huffaker & Robbins 1982).

[105]

The library we used was constructed by Carlson & Botstein (1982) and consists of YEp24 plasmids (Botstein *et al.* 1979) into which random fragments of strain DBY939 yeast genomic DNA of size 10 kilobases or larger were inserted. YEp24 (figure 7) transforms both yeast and *Escherichia coli*, replicates episomally, and allows for the expression of any gene in the insert. It contains the complete yeast *URA3* gene, which is used for the selection of transformants in yeast ura3⁻ strains. *E. coli* transformants can be selected by using the *tet*ʳ and *amp*ʳ drug resistance markers or alternatively by complementing the *pyrF* mutation with the *URA3* gene.

TABLE 1. PLASMID SEGREGATION IN YEAST TRANSFORMANTS

	platings from Alg1 selective growth (36 °C, +U)		platings from non-selective growth (26 °C, +U)	
	growth ratio in plates with U +U, 36 °C/26 °C	growth ratio in plates at 26 °C 26 °C, ⁻U/+U	growth ratio in plates with U +U, 36 °C/26 °C	growth ratio in plates at 26 °C 26 °C, ⁻U/+U
1	0.61	0.56	0.05	0.05
2	0.61	0.59	0.06	0.05
3	0.58	0.52	0.08	0.09
4	0.81	0.71	0.36	0.21
5	0.79	0.72	0.40	0.40
alg⁺ ura⁻	—	0.00	—	—

Transformants were grown for 2 days in complete liquid medium either with selective pressure for Alg1 (36 °C) but not for Ura3, or non-selectively for both markers. They were then plated either at 26 °C or at 36 °C, with or without uracil. The ratios from these platings were calculated after averaging from three plates in each particular case. The last line in the table shows the results of a control experiment in which the strain *ALG1⁺ura3⁺*, transformed with YEp24, was grown non-selectively in complete medium at 36 °C.

We prepared library plasmid DNA and used it to transform the yeast double mutant *alg1-1, ura3-52*. Selection of Alg1⁺Ura3⁺-complemented transformants was done in two stages. In the first stage we recovered yeast transformants growing in medium without uracil at 26 °C, the permissive temperature for *alg1-1*. In the second stage the Ura3⁺ transformants were screened for growth at 36 °C, the non-permissive temperature for *alg1-1*. The transformants that could grow at 36 °C in medium lacking uracil were then subjected to plasmid segregation experiments to eliminate the possibility that either the *ura3-52* or the *alg1-1* mutation had reverted, and to show that both mutations were being complemented by the same plasmid. We showed that when we removed the selective pressure for both markers, by growing at 26 °C in complete medium, the transformants lost both the Alg1⁺ and the Ura3⁺ phenotypes (table 1). However, when we maintained selective pressure on Alg1 but not on Ura3, by growing at 36 °C in complete medium, the transformants retained both phenotypes (table 1). We therefore concluded that the yeast transformants contained a plasmid capable of complementing the *alg1-1* and the *ura3-52* mutations simultaneously.

We transformed *E. coli* HB101 to ampicillin resistance with a total DNA extract of yeast transformants. We then isolated the plasmid pJC1. Restriction analysis showed that pJC1 was YEp24 ligated to an 11 kilobase insert at the *Bam*HI site. This plasmid could transform the yeast double mutant *alg1-1ura3-52* and complement both mutations.

We are now subcloning the yeast DNA insert in pJC1 and are preparing to clone other *alg* mutations. After we prove that we have a series of genes in the asparagine-linked glycosylation pathway, we plan to study structural features of the gene products, their regulation, synthesis and mode of action. In addition, increased gene copy number should lead to an overproduction of the enzyme and thus will facilitate its isolation and characterization.

TOPOGRAPHY OF LIPID-LINKED OLIGOSACCHARIDE SYNTHESIS

The assembly of the lipid-linked oligosaccharide and its transfer to protein involve the transmembrane movement of sugar residues. Oligosaccharides of newly made glycoproteins are first found segregated within the lumen of the e.r. This segregation within the e.r. is the first step in the secretion process. Because these oligosaccharides are assembled from cytoplasmic sugar nucleotides (Coates *et al.* 1980), it is clear that the synthesis of the oligosaccharide-lipid and its transfer to protein are coupled to the transport of sugar residues across the e.r. membrane. Over the last several years, we have been studying how this transmembrane movement occurs.

Our experiments have used microsomal vesicles *in vitro*. These are sealed vesicles derived from rough and smooth e.r. Microsomes have a unique orientation, with the cytoplasmic side of the membrane facing the medium. Thus non-penetrating reagents will act only on the cytoplasmic side of the membrane. We and others have used this system to examine the transmembrane orientation of oligosaccharide-lipids, and the enzymes responsible for their synthesis.

Studies of enzymes of oligosaccharide–lipid synthesis

We have used proteases as probes to study the orientation of enzymes of oligosaccharide assembly (Snider *et al.* 1980). Microsomes from rat liver were used. Figure 8 shows an experiment in which intact vesicles and vesicles made leaky with a non-ionic detergent were treated with pronase. Enzymes of oligosaccharide-lipid synthesis and the microsomal enzyme glucose-6-phosphatase were then assayed. As previously observed, glucose-6-phosphatase, an enzyme of the microsomal lumen, was not inactivated by pronase treatment of intact vesicles. However, in leaky vesicles, most of this activity was lost. Thus pronase did not penetrate intact microsomal vesicles under our conditions of treatment.

In contrast, all the activities of oligosaccharide-lipid synthesis that we assayed were inactivated by pronase treatment of intact vesicles. Moreover, the protease sensitivity of the enzymes was the same in intact and leaky vesicles, suggesting that the protease-sensitive sites of these enzymes are on the cytoplasmic side of the microsomal membrane.

We found that the synthesis of Dol-P-Man and Dol-P-Glc were largely inactivated by pronase treatment of intact microsomes (figure 8). Activity in the other two assays was less sensitive to pronase. This can be explained by the fact that these latter assays measure groups of reactions; one reaction in each group was not protease sensitive under our conditions of treatment. This was demonstrated by fractionating the products of these reactions. In the assay measuring the incorporation of GlcNAc residues from UDP-GlcNAc into lipid, we showed that the synthesis of $GlcNAc_2$-lipid is protease sensitive in both intact and leaky vesicles. However, GlcNAc-lipid synthesis is not protease sensitive under either condition. Similarly, in the assay measuring the incorporation of Glc residues from Dol-P-Glc into oligosaccharide-lipid, protease treatment of intact microsomes resulted in large decreases in the synthesis of $Glc_3Man_9GlcNAc_2$-lipid and $Glc_2Man_9GlcNAc_2$-lipid, whereas the synthesis of the monoglucosyl species was largely unaffected.

These results show that at least four reactions of oligosaccharide-lipid synthesis have protease-sensitive sites on the cytoplasmic side of the microsomal membrane. In addition, none of the enzymes we examined were protease-resistant in intact vesicles but protease-sensitive in leaky vesicles. Thus no enzymes had protease-sensitive sites exclusively on the luminal side of the membrane. How are these enzymes oriented in the microsomal membrane? These results are consistent with two possible arrangements. The first is that of enzymes with catalytic sites facing the cytoplasm. In the second, transmembrane enzymes have luminal catalytic sites, but cytoplasmic protease-sensitive sites. If this second possibility is correct, it is likely that the cytoplasmic segments of these enzymes play an important role in the function of these enzymes. This argument is based on the finding that five other enzymes of the e.r. and Golgi apparatus that have luminal active sites are protease resistant in intact vesicles, but sensitive in leaky vesicles (DePierre & Dallner 1975; Grinna & Robbins 1979; Carey & Hirschberg 1981; Fleischer 1981). No enzymes showed the behaviour that we found for the enzymes of oligosaccharide-lipid synthesis.

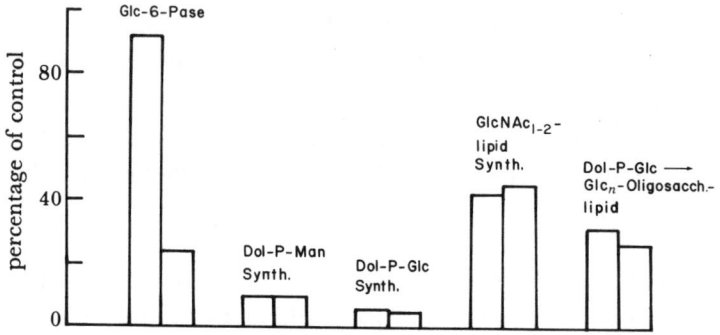

FIGURE 8. Enzymes of oligosaccharide-lipid synthesis have protease-sensitive sites on the cytoplasmic side of the microsomal membrane. Rat liver microsomal vesicles were treated with pronase, and enzymes of oligosaccharide-lipid synthesis and glucose-6-phosphatase were assayed as previously described (Snider *et al.* 1980). Results from treatment of intact vesicles (left-hand bars) and leaky vesicles (right-hand bars) are expressed as a percentage of control samples that were not treated with protease.

Both of these possible arrangements of the synthetic enzymes argue strongly that sugar nucleotides are utilized directly at the cytoplasmic face of the e.r. during oligosaccharide-lipid synthesis. Supporting this conclusion is the recent demonstration by Hanover & Lennarz (1982) that UDP-GlcNAc, GDP-Man and UDP-Glc are not able to penetrate microsomal vesicles *in vitro*. This finding is significant in light of the fact that sugar nucleotide transport *has* been observed in Golgi vesicles by using similar methods (Carey *et al.* 1980).

Orientation of mature oligosaccharide-lipid

Although sugar nucleotides are utilized at the cytoplasmic face of the e.r. membrane, it is clear that sugars must cross this membrane during the assembly of oligosaccharide-lipid. This conclusion is based on our recent demonstration that the largest oligosaccharide-lipid is found on the luminal side of the membrane (Snider & Robbins 1982). These studies used concanavalin A (Con A), a plant lectin that binds tightly to the precursor oligosaccharide-lipid. In order to use this lectin as a probe for the orientation of oligosaccharide-lipid, we developed an assay for the binding of lectin to this compound. The assay is based on the solubility of the Con A–oligosaccharide-lipid complex. Although oligosaccharide-lipid is extracted with chloroform–methanol–water (10:10:3), the Con A–oligosaccharide-lipid complex is not. Thus Con A binding is measured as a loss of extractable oligosaccharide-lipid.

We have examined the binding of Con A to oligosaccharide-lipid in microsomes prepared from [³H]mannose-labelled cultured fibroblasts (figure 9). In intact vesicles, a small fraction of the oligosaccharide-lipid was bound by Con A. In contrast, most of this material was bound by

Con A in leaky vesicles. In addition, much of the binding seen in intact vesicles was probably due to the fact that roughly 20 % of the vesicles in the original preparation were leaky. This approach was extended by fractionating the lipid-linked oligosaccharides in these samples. In control vesicles, the major lipid-linked species was the full-sized oligosaccharide Glc_3Man_9-$GlcNAc_2$. In intact vesicles, only 12 % of this compound was bound by Con A, while in leaky vesicles 75 % was bound by Con A (not shown).

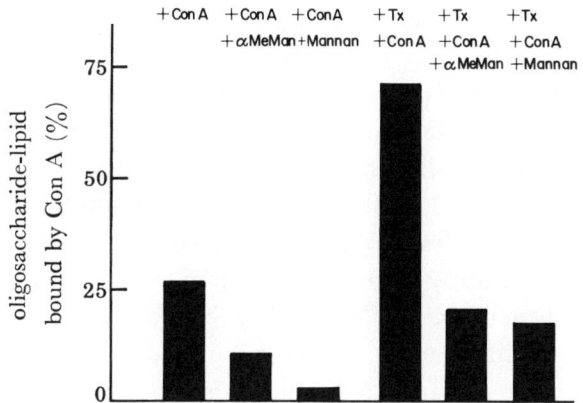

FIGURE 9. Oligosaccharide-lipid is located in the luminal side of microsomal vesicles. Membranes prepared from [^{3}H]mannose-labelled Chinese hamster ovary cell cultures were incubated with Con A and, where indicated, Triton X-100 (Tx), α-methyl mannoside (αMeMan) or yeast mannan. The amount of oligosaccharide-lipid bound by Con A was determined as described in the text (Snider & Robbins 1982). The amount of oligosaccharide-lipid bound by Con A is much greater in the presence of detergent, suggesting that the compound faces the luminal side of the vesicles. The specificity of Con A binding oligosaccharide-lipid is shown by the fact that it is inhibited by the competitors α-methyl mannoside and mannan.

These results suggest that the largest lipid-linked oligosaccharide, the donor in protein glycosylation, faces the microsomal lumen. Thus, sugar residues must cross the microsomal membrane during the assembly of this compound. Moreover, this orientation argues that oligosaccharide is transferred from lipid to protein on this side of the membrane. This conclusion is supported by several lines of evidence. First, oligosaccharide transferred from lipid to peptide acceptors was found within the lumen of hen oviduct microsomes *in vitro*, even though peptide translocation probably did not occur (Hanover & Lennarz 1980). Moreover, glycoprotein polypeptides translated *in vitro* were glycosylated only if they were cotranslationally inserted into microsomal membranes (Katz *et al.* 1977; Lingappa *et al.* 1978). Second, the precursor to adrenocorticotrophic hormone is glycosylated post-translationally in adrenal tumour cells *in vivo*, presumably while the polypeptide is sequestered in the e.r. lumen (Roberts *et al.* 1978). Finally, a study on the glycosylation of ovalbumin nascent chains suggested that glycosylation could occur only after there was a peptide segment between the ribosome and the acceptor site on the peptide chain long enough to span the bilayer (Glabe *et al.* 1981).

How then do sugar residues derived from cytoplasmic sugar nucleotides move across the e.r. membrane during the assembly of luminal oligosaccharide-lipid? One possibility is the coupling of glycosylation to transport: glycosyltransferases using cytoplasmic sugar nucleotides would carry residues across the membrane and add them to growing oligosaccharide chains on the luminal side of the membrane. Such a mechanism is likely for the synthesis of the second intermediate in the pathway, $Dol-PP-GlcNAc_2$. Hanover & Lennarz (1979, 1982) have localized this compound to the luminal side of microsomal vesicles. In addition, those workers could find no evidence for the appearance of this compound on the cytoplasmic side of the membrane during its synthesis. However, the assembly of this compound was sensitive to proteolysis from the cytoplasmic side of the membrane (Snider *et al.* 1980; Hanover & Lennarz 1982). Because neither UDP-GlcNAc nor the intermediate itself can cross the membrane, it is likely that the transport of GlcNAc residues occurs as an integral part of the glycosylation process.

A second possibility is that sugar residues are transported by the transmembrane movement of the lipid-linked intermediates themselves. This possibility has been ruled out for two dolichyl-pyrophosphate-linked oligosaccharides, $GlcNAc_2$-lipid (Hanover & Lennarz 1979, 1982) and $Glc_3Man_9GlcNAc_2$-lipid (Snider & Robbins 1982). However, the recent studies of Haselbeck & Tanner (1982) suggest that Dol-P-Man, an intermediate that donates the outer four mannose residues to growing oligosaccharide-lipids, may serve to transport mannose residues into the lumen. Those workers have presented evidence for the transmembrane movement of Dol-P-Man in reconstituted phospholipid vesicles containing purified Dol-P-Man synthetase from yeast. This translocation is almost certainly mediated by protein because similar compounds do not flip spontaneously in model membranes (McCloskey & Troy 1980). If the same translocation occurs *in vivo*, Dol-P-Man might be made on the cytoplasmic side of the e.r. membrane, and then flip so that it can serve as a mannose donor in the assembly of luminal oligosaccharide-lipid. However, confirmation of this model will require the direct probing of the orientation of this compound in the e.r. membrane.

These studies have greatly increased our understanding of the transport of sugar residues that occurs during the synthesis of asparagine-linked oligosaccharides. The demonstration that nucleotide sugars are utilized from the cytoplasmic side of the membrane allows the possibility of uptake into the lumen to be ruled out. Similarly, the luminal location of mature oligo-saccharide-lipid and $GlcNAc_2$-lipid rules out the possibility of assembly of the oligosaccharide on the cytoplasmic face of the membrane. These studies have also suggested two possible transport mechanisms. The first, which suggests the coupling of glycosylation and transport, could be active in the addition of the first seven sugar residues of the oligosaccharide-lipid, which are added directly from sugar nucleotides. The second possible transport mechanism, involving the flipping of Dol-P-Man, could function in the addition of the outer sugars of the oligo-saccharide-lipid. The confirmation of these mechanisms, as well as the details of the transport processes, remain a subject for future experiments.

CONTROL OF ASPARAGINE-LINKED OLIGOSACCHARIDE PROCESSING

Immediately after transfer to proteins, the initially homogeneous population of precursor oligosaccharides (compound *a*, figure 10) begins to undergo a series of modifications that will eventually produce the diverse array of *N*-linked glycans found on mature glycoproteins. We have previously shown that in rat liver, two enzymes of the e.r. lumen, glucosidases I and II, sequentially remove the three Glc residues to produce $Man_9GlcNAc_2$ (compound *b*, figure 10) (Grinna & Robbins 1979). Two Golgi α-mannosidases (mannosidases I–A and I–B) can then remove up to four of the nine Man residues to produce $Man_5GlcNAc_2$ (compound *c*, figure 10) (Tabas & Kornfeld 1979; Tulsiani *et al.* 1982). These reactions yield $Man_{5-9}GlcNAc_2$ structures, termed high-mannose.

The synthesis of complex-type oligosaccharides, a second major class of structures, begins with the action of GlcNAc transferase I on $Man_5GlcNAc_2$ to yield compound d in figure 10. After the removal of two more mannose residues, mature complex structures (compounds f and g) are formed by the addition of a variety of peripheral sugar residues.

Although the outlines of this processing pathway are now clear, a key question remains: what factors determine whether a given glycosylation site will carry an extensively processed (i.e. complex) or partly processed (i.e. high-mannose) oligosaccharide? We have been investigating this problem by using the two membrane glycoproteins of Sindbis virus, E1 and E2. Because these glycoproteins have both high-mannose and complex oligosaccharides, they represent an excellent system for investigating this problem.

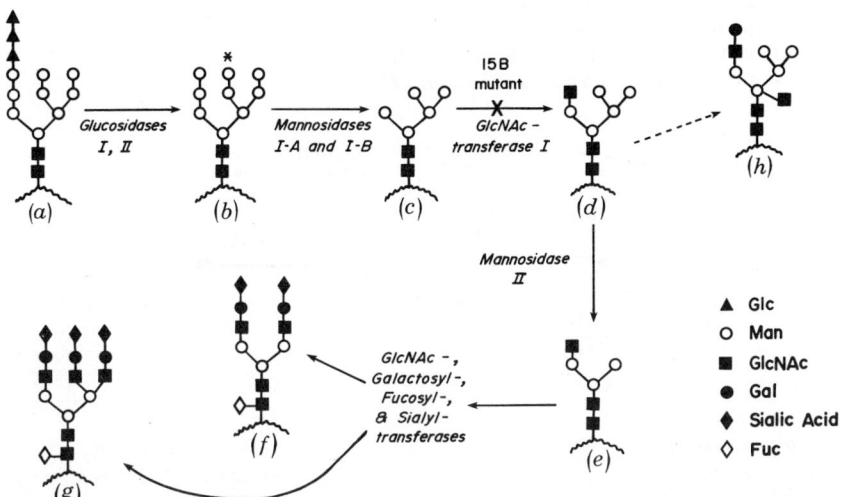

FIGURE 10. Processing of asparagine-linked oligosaccharides. The conversion of the $Glc_3Man_9GlcNAc_2$ precursor oligosaccharide (a) to high-mannose (b, c) and complex (f, g) oligosaccharides is shown.

Correlation of extent of processing with 'exposure' of oligosaccharide site measured with glycosidase probes

Although the requirement for an -Asn-X-Ser/Thr- sequence in N-glycosylation is well established, no specific amino acid sequence directing oligosaccharide processing to high-mannose or complex-type structures has been found. This has led us to the hypothesis that oligosaccharide processing enzymes are relatively insensitive to detailed protein structure but act on any appropriate oligosaccharide structure sterically available to their active sites. Thus, the N-linked glycans occupying the most 'exposed' sites on the surface of a native glycoprotein would be expected to have fully processed (complex-type) structures, whereas those obscured by polypeptide chain folding or other interactions would retain high-mannose structures.

The evaluation of this hypothesis has necessitated the development of techniques to separate individual glycosylation sites to allow the examination of their oligosaccharide structures. This is accomplished by high-performance liquid chromatography (h.p.l.c.) of tryptic glycopeptides. Using this technique we have characterized four sites on E1 and E2 as either complex, high-mannose or variable (see below). The relative 'exposure' of glycosylation sites on the surface of a native glycoprotein was determined by using the enzyme endo-β-N-acetylglucosaminidase H (endo H), which cleaves between the two GlcNAc residues in high-mannose oligosaccharides.

To use this enzyme as a probe for the exposure of complex sites as well as high-mannose sites, we performed these experiments on the glycoproteins of Sindbis virions grown in a mutant cell line (15B) in which sites normally carrying complex oligosaccharides carry $Man_5GlcNAc_2$ instead.

The experiment is presented in schematic form and results are shown in figure 11 (Hsieh & Robbins 1982). In excellent agreement with the hypothesis, endo H released oligosaccharides from the two complex sites more rapidly than from the two high-mannose sites. All four sites were equally susceptible to endo H after denaturation or pronase digestion.

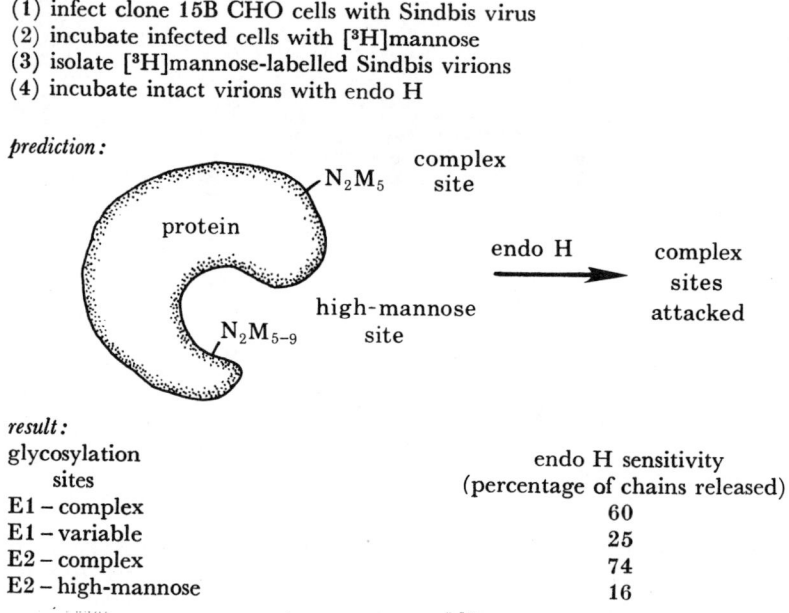

experiment:
(1) infect clone 15B CHO cells with Sindbis virus
(2) incubate infected cells with [³H]mannose
(3) isolate [³H]mannose-labelled Sindbis virions
(4) incubate intact virions with endo H

prediction:

result:

glycosylation sites	endo H sensitivity (percentage of chains released)
E1 – complex	60
E1 – variable	25
E2 – complex	74
E2 – high-mannose	16

FIGURE 11. Oligosaccharides at complex glycosylation sites are more exposed than oligosaccharides at high-mannose or variable sites in Sindbis virus glycoproteins. Sindbis virions were isolated from [³H]mannose-labelled infected CHO clone 15B cells. Intact virions were then treated with endo H under native conditions. The degree of cleavage by endo H at each site was analysed by h.p.l.c. of tryptic glycopeptides from treated and untreated samples (Hsieh & Robbins 1982). The sites that normally bear complex oligosaccharides are much more susceptible to endo H cleavage than the high-mannose site or the variable site (see text).

Related experiments with endo H as a probe have been carried out with yeast invertase by R. B. Trimble, and with mouse β-glucuronidase by J. U. Baenziger (personal communication). Trimble has found that the large mannan-type oligosaccharide chains are much more susceptible to endo H cleavage in native invertase than 'core' glycosylation sites, which carry smaller high-mannose chains of about 12–16 sugar residues. Baenziger has demonstrated endo H sensitivity for the phosphorylated (processed) oligosaccharides of β-glucuronidase and endo H resistance for unphosphorylated residues. In each case both types of oligosaccharide were sensitive to endo H action after protein denaturation.

In addition to the general correlation between processing and 'exposure' discovered above, we have recently observed a striking variability in the ability of different host cells to process particular glycosylation sites. The gel filtration profile of glycopeptides and oligosaccharides attached to one of the glycosylation sites of E1 after growth of the virus in three separate hosts is shown in figure 12. Clearly, in chick embryo fibroblasts this site is left as a high-mannose structure (figure 12*a*). In contrast, the same site is extensively processed to complex oligosaccharides in BHK cells (figure 12*b*) and becomes a mixture of complex and high-mannose oligosaccharides when the virus is grown in CHO cells (figure 12*c*).

It is conceivable that oligosaccharides at a particular site may be oriented such that processing enzymes of one cell can act extensively at the site whereas enzymes from another cell are less

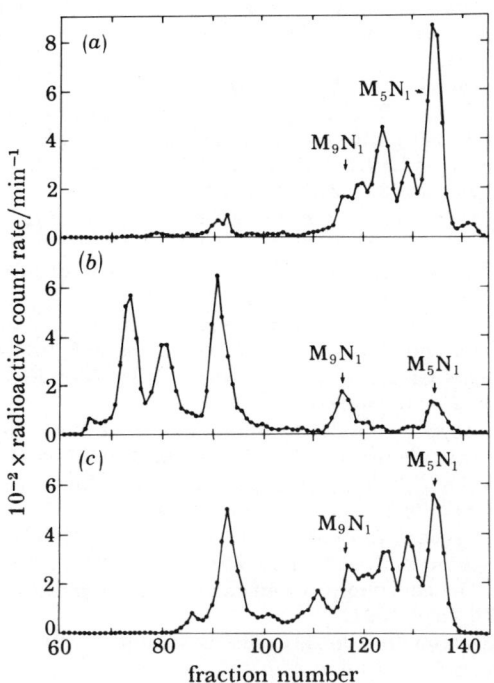

FIGURE 12. Variability with host cell type of the oligosaccharides at one of the glycosylation sites of glycoprotein E1. Labelled Sindbis virions were prepared from [³H]mannose-labelled infected cells, and the tryptic glycopeptides representing one of the glycosylation sites of E1 prepared by h.p.l.c. Samples were then treated with pronase and endo H. The resulting glycopeptides and oligosaccharides were analysed by gel filtration chromatography on Bio-Gel P-4 (minus 400 mesh) (Hsieh & Robbins 1982). Complex glycopeptides elute in the left half of each profile, while high-mannose oligosaccharides are found in the right half. The elution positions of Man₉GlcNAc (M₉N₁) and Man₅GlcNAc (M₅N₁) are indicated. Profiles for this E1 site from virions grown in (*a*) primary chick embryo fibroblasts, (*b*) BHK-21 cells and (*c*) CHO cells are shown.

able to do so. Alternatively, different cells may have varying effects on the nature or duration of physical availability of oligosaccharides at a single site to processing enzymes. Either of these possibilities would account for host-dependent variation seen in this site on E1. It remains to be determined whether this same phenomenon also occurs with other glycoproteins.

We thank R. Das, P. Hsieh, S. C. Hubbard, A. Mercurio and K. Runge for helpful discussions. This work was supported by grants no. CA14142 and CA14051 (to P.W.R. and S.E.L.) from the National Cancer Institute, Department of Health, Education and Welfare.

15-2

REFERENCES

Ballou, C. E. 1980 In *Fungal polysaccharides* (ed. P. A. Sanford & K. Matsuda), pp. 1–14. Washington, D.C.: American Chemical Society.

Botstein, D., Falco, S. C., Stewart, S. E., Brennan, M., Scherer, S., Stincomb, D. T., Struhl, K. & Davis, R. W. 1979 Sterile host yeasts (SHY): a eukaryotic system of biological containment for recombinant DNA experiments. *Gene* **8**, 17–23.

Carey, D. J. & Hirschberg, C. B. 1981 Topography of sialoglycoproteins and sialyltransferases in mouse and rat liver Golgi. *J. biol. Chem.* **256**, 989–993.

Carey, D. J., Sommers, L. W. & Hirschberg, C. B. 1980 CMP-*N*-acetylneuraminic acid: isolation from and penetration into mouse liver microsomes. *Cell* **19**, 597–605.

Carlson, M. & Botstein, D. 1982 Two differentially regulated mRNAs with different 5′ ends encode secreted and intracellular forms of yeast invertase. *Cell* **28**, 145–154.

Chapman, A., Trowbridge, I. S., Hyman, R. & Kornfeld, S. 1979 Structure of the lipid-linked oligosaccharides that accumulate in Class E Thy-1-negative mutant lymphomas. *Cell* **17**, 509–515.

Coates, S. W., Gurney, T. Jr, Sommers, L. W., Yeh, M. & Hirschberg, C. B. 1980 Subcellular localization of sugar nucleotide synthetases. *J. biol. Chem.* **255**, 9225–9229.

DePierre, J. W. & Dallner, G. 1975 Structural aspects of the membrane of the endoplasmic reticulum. *Biochim. biophys. Acta* **415**, 411–472.

Fleischer, B. 1981 Orientation of glycoprotein galactosyltransferase and sialyltransferase enzymes in vesicles derived from rat liver Golgi apparatus. *J. Cell Biol.* **89**, 246–255.

Glabe, C. G., Hanover, J. A. & Lennarz, W. J. 1980 Glycosylation of ovalbumin nascent chains. The spatial relationship between translation and glycosylation. *J. biol. Chem.* **255**, 9236–9242.

Grinna, L. S. & Robbins, P. W. 1979 Glycoprotein biosynthesis. Rat liver microsomal glucosidases which process oligosaccharides. *J. biol. Chem.* **254**, 8814–8818.

Hanover, J. A. & Lennarz, W. J. 1979 The topological orientation of *N,N*′-diacetylchitobiosylpyrophosphoryldolichol in artificial and natural membranes. *J. biol. Chem.* **254**, 9237–9246.

Hanover, J. A. & Lennarz, W. J. 1980 *N*-linked glycoprotein assembly. Evidence that oligosaccharide attachment occurs within the lumen of the endoplasmic reticulum. *J. biol. Chem.* **255**, 3600–3604.

Hanover, J. A. & Lennarz, W. J. 1982 Transmembrane assembly of *N*-linked glycoproteins. Studies on the topology of saccharide-lipid synthesis. *J. biol. Chem.* **257**, 2787–2794.

Haselbeck, A. & Tanner, W. 1982 Dolichyl phosphate-mediated mannosyl transfer through liposomal membranes. *Proc. natn. Acad. Sci. U.S.A.* **79**, 1520–1524.

Hasilik, A. & Tanner, W. 1978 Carbohydrate moiety of carboxypeptidase Y and perturbation of its biosynthesis. *Eur. J. Biochem.* **91**, 567–575.

Hsieh, P. & Robbins, P. W. 1982 (In preparation.)

Hubbard, S. C. & Ivatt, R. J. 1978 Synthesis and processing of asparagine-linked oligosaccharides. *A. Rev. Biochem.* **50**, 555–583.

Huffaker, T. C. & Robbins, P. W. 1982 Temperature-sensitive yeast mutants deficient in asparagine-linked glycosylation. *J. biol. Chem.* **257**, 3203–3210.

Hunt, L. A. 1980 CHO cells selected for phytohemagglutinin and Con A resistance are defective in both early and late stages of protein glycosylation. *Cell* **21**, 407–415.

Katz, F. N., Rothman, J. E., Lingappa, V. R., Blobel, G. & Lodish, H. F. 1977 Membrane assembly *in vitro*: synthesis, glycosylation, and asymmetric insertion of a transmembrane protein. *Proc. natn. Acad. Sci. U.S.A.* **74**, 3278–3282.

Lehle, L. 1980 Biosynthesis of the core region of yeast mannoproteins. *Eur. J. Biochem.* **109**, 589–601.

Lehle, L., Cohen, R. E. & Ballou, C. E. 1979 Carbohydrate structure of yeast invertase. *J. biol. Chem.* **254**, 12209–12218.

Lingappa, V. R., Lingappa, J. R., Prasad, R., Ebner, K. E. & Blobel, G. 1978 Coupled cell-free synthesis, segregation and core glycosylation of a secretory protein. *Proc. natn. Acad. Sci. U.S.A.* **75**, 2338–2342.

McCloskey, M. A. & Troy, F. A. 1980 Paramagnetic isoprenoid carrier lipids. 2. Dispersion and dynamics in lipid membranes. *Biochemistry, Wash.* **19**, 2061–2066.

Maniatis, T., Ross, C. H. & Lacy, E. 1978 The isolation of structural genes from libraries of eucaryotic DNA. *Cell* **15**, 687–701.

Meyer, J. & Matile, P. 1974 Regulation of isoenzymes and secretion of invertase in bakers' yeast. *Biochem. Physiol. Pflanzen* **166**, 377–385.

Parodi, A. J. 1979 Biosynthesis of yeast mannoproteins. *J. biol. Chem.* **254**, 10051–10060.

Perlman, D. & Halvorson, H. O. 1981 Distinct repressible mRNAs for cytoplasmic and secreted yeast invertase are encoded by a single gene. *Cell* **25**, 525–536.

Roberts, J. L., Phillips, M., Rosa, P. A. & Herbert, E. 1978 Steps involved in the processing of common precursor forms of adrenocorticotropin and endorphin in cultures of mouse pituitary cells. *Biochemistry, Wash.* **17**, 3609–3618.

Snider, M. D. & Robbins, P. W. 1982 Transmembrane organization of protein glycosylation. Mature oligo-saccharide-lipid is located on the luminal side of microsomes from Chinese hamster ovary cells. *J. biol. Chem.* **257**, 6796–6801.

Snider, M. D., Sultzman, L. A. & Robbins, P. W. 1980 Transmembrane location of oligosaccharide-lipid synthesis in microsomal vesicles. *Cell* **21**, 385–392.

Stanley, P. 1980 In *The biochemistry of glycoproteins and proteoglycans* (ed. W. Lennarz), pp. 161–189. New York: Plenum Press.

Struck, D. K. & Lennarz, W. J. 1980 In *The biochemistry of glycoproteins and proteoglycans* (ed. W. Lennarz), pp. 35–83. New York: Plenum Press.

Tabas, I. & Kornfeld, S. 1979 Purification of a rat liver golgi α-mannosidase capable of processing asparagine-linked oligosaccharides. *J. biol. Chem.* **254**, 11655–11663.

Tarentino, A. L., Plummer, T. H. Jr & Maley, F. 1974 The release of intact oligosaccharides from specific glycoproteins by endo-β-N-acetylglucosaminidase H. *J. biol. Chem.* **249**, 818–824.

Trimble, R. B. & Maley, F. 1977 The use of endo-β-N-acetylglucosaminidase H in characterizing the structure and function of glycoproteins. *Biochem. biophys. Res. Commun.* **78**, 935–944.

Trimble, R. B., Maley, F. & Tarentino, A. L. 1980 Characterization of large oligosaccharide-lipids synthesized *in vitro* by microsomes from *Saccharomyces cerevisiae*. *J. biol. Chem.* **255**, 10232–10238.

Tulsiani, D. R. P., Hubbard, S. C., Robbins, P. W. & Touster, O. 1982 α-D-Mannosidases of rat liver Golgi membranes. Mannosidase II is the GlcNAcMan$_5$-cleaving enzyme in glycoprotein biosynthesis and mannosidases IA and IB are the enzymes converting Man$_9$ precursors to Man$_5$ intermediates. *J. biol. Chem.* **257**, 3660–3668.

Discussion

T. Feizi (*Division of Communicable Diseases, Clinical Research Centre, Harrow, U.K.*). One possible explanation for differences in degree of processing of oligosaccharides at the same glycosylation site is that the cells in a given cell line are not absolutely uniform and the glycoproteins are a mixture of products derived from cells that are subtly different. In that case (and if these differences are stable) it might be possible to clone the cells and obtain glycoproteins with uniform glycosylation. We often find heterogeneity of the expression of carbohydrate antigens on cultured cell lines (Childs *et al.* 1980).

Reference

Childs, R. A., Kapadia, A. & Feizi, T. 1980 Expression of blood group I and i active carbohydrate sequences on cultured human and animal cell lines assessed by radioimmunoassays with monoclonal cold agglutinins. *Eur. J. Immunol.* **10**, 379–384.

Phil. Trans. R. Soc. Lond. B **300**, 225–228 (1982)

Printed in Great Britain

The protein translocation machinery of the endoplasmic reticulum

BY P. WALTER, R. GILMORE, M. MÜLLER AND G. BLOBEL

Laboratory of Cell Biology, The Rockefeller University, New York, NY 10021, U.S.A.

The rough endoplasmic reticulum (r.e.r.) has been postulated to possess a single translation-coupled translocation system (in multiple copies) that effects signal sequence-mediated translocation of all secretory and lysosomal proteins and integration of all integral membrane proteins whose port of entry is the rough endoplasmic reticulum (G. Blobel 1980 *Proc. natn. Acad. Sci. U.S.A.* **77**, 1496–1500). Two proteins have been isolated that are components of the r.e.r. translocation system. Their properties and function in protein translocation across and integration into membranes are discussed.

Substantial experimental data have recently been provided on the co-translational translocation of proteins across and integration into the endoplasmic reticulum. So far, two components have been purified from dog pancreas and shown to be required for this translocation process.

One of these is the so-called signal recognition particle (SRP), an $11S$ ribonucleoprotein (Walter & Blobel 1982a). SRP consists of six non-identical polypeptide chains (molecular masses 72, 68, 54, 19, 14 and 9 kDa) (Walter & Blobel 1980) and one molecule of $7S$ RNA (Walter & Blobel 1982a). The RNA has been identified by partial sequence analysis (Walter & Blobel 1982a) to be the previously described (Zieve & Penman 1976) and recently sequenced (Ullu *et al.* 1982; Li *et al.* 1982) small cytoplasmic $7S$ RNA ($7S$ RNA, ScL). Both RNA and protein are required for SRP's activity. In dog pancreas at physiological salt concentration (150 mM potassium ions) the bulk of SRP appears to be about equally distributed between a membrane-bound and a free or ribosome/polysome-associated form (Walter & Blobel 1982b).

The other component, termed SRP receptor (Gilmore *et al.* 1982a), is a protein of molecular mass 72 kDa (Gilmore *et al.* 1982b; Meyer *et al.* 1982b) that has been purified from detergent-solubilized microsomal membranes by SRP-affinity chromatography (Gilmore 1982b). The SRP receptor is an integral membrane protein of the endoplasmic reticulum. It consists of a large cytoplasmic domain of molecular mass 60 kDa (Meyer & Dobberstein 1980b) that can be severed from the membrane in an intact form by treatment with a variety of proteases and can be added back to the proteolysed membranes to reconstitute activity (Gilmore *et al.* 1982a; Walter *et al.* 1979; Meyer & Dobberstein 1980a).

The function of these components in the protein translocation process was deduced from assay systems reconstituted *in vitro*. By using such assays, SRP was found to function in decoding the information contained in the signal peptide of nascent secretory (Walter *et al.* 1981; Stoffel *et al.* 1981; Muller *et al.* 1982), lysosomal (Erickson *et al.* 1982) and membrane (Anderson *et al.* 1982) proteins to the effect that it mediates the specific attachment of the translating ribosome to the microsomal membrane (Walter & Blobel 1981a). In the absence of microsomal membranes SRP specifically arrests the elongation of secretory protein synthesis *in vitro* (Walter *et al.* 1981) just after the signal peptide has emerged from the ribosome, thereby preventing the completion of pre-secretory proteins (many of which may be potentially harmful to the cell) (Walter & Blobel 1981 *b*) in the cytoplasmic compartment. Upon interaction of these arrested ribosomes with a specific integral membrane protein, the SRP receptor (Gilmore *et al.* 1982 *a*; Meyer *et al.* 1982 *a*), on the microsomal membrane, this elongation arrest is released and the nascent chain is translocated across (Walter & Blobel 1981 *b*) or – as in integral membrane proteins – integrated into (Anderson *et al.* 1982) the lipid bilayer.

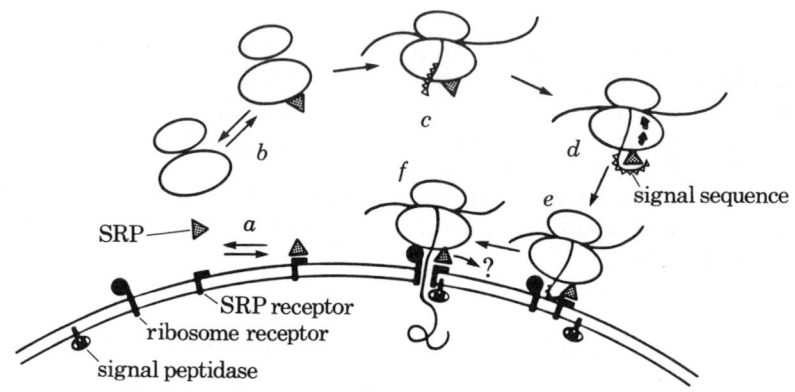

FIGURE 1. Model for co-translational protein translocation across the rough endoplasmic reticulum membrane. For details see text.

The drawing in figure 1 represents a model (taken from Walter & Blobel 1981 *b*) illustrating schematically both facts and speculations about protein translocation across the rough endoplasmic reticulum (r.e.r.). It was proposed (Walter & Blobel 1981 *b*, 1982 *b*) that an equilibrium exists between a free, soluble form of SRP, SRP bound to ribosomes and SRP bound to the SRP receptor (figure 1*a*, *b*). Upon translation of an mRNA coding for a signal sequence (figure 1*c*) that is addressed to the r.e.r. translocation system and that is present in all secretory proteins, all lysosomal proteins and all those integral membrane proteins whose exclusive site of integration is the r.e.r., there is an enhancement of the apparent affinity of SRP for the translating ribosomes by several orders of magnitude (figure 1*d*). Concomitantly, and presumably through the ribosome, SRP arrests the elongation of the initiated polypeptide chain, preventing its completion in the cytoplasm. Translation arrest is released only upon interaction of the SRP arrested ribosome with the SRP receptor (figure 1*e*).

We have estimated (Gilmore *et al.* 1982 *b*; Walter *et al.* 1981) that one equivalent of dog pancreas microsomal membranes contains approximately 500 fmol of bound ribosomes, approximately 20 fmol of SRP, and about 100 fmol of SRP receptor. Thus the content of both SRP and SRP receptor is less than that of bound ribosomes. This suggested (Walter & Blobel 1981 *b*) that the ribosome–SRP–SRP-receptor interaction might be a transient one, merely targeting the SRP-arrested ribosome to a specific membrane site that is represented in part by the SRP receptor and in part by other integral membrane proteins. The latter could be represented by ribophorins I and II, which have been characterized by Kreibich & Sabatini and their coworkers (Kreibich *et al.* 1978 *a*, *b*; Marcantonio *et al.* 1982). Once targeting has occurred, the ribosome–SRP–SRP-receptor interaction might then be replaced (figure 1*f*) by a direct interaction of the ribosomes with ribophorins I and II, an interaction that might persist for the entire chain translocation event. Ribophorins I and II have been reported each to be present in microsomal membranes in stoichiometric amounts to those of membrane-bound ribosomes (Kreibich *et al.* 1978*a*, *b*). It should be noted, however, that Bielinska *et al.* (1979) have argued. against the involvement of ribophorins in chain translocation. In any case it should be emphasized that these proposals suggesting a possible cascade in the formation of a productive ribosome–membrane junction and involving several integral membrane proteins are at this moment entirely speculative.

The ability of SRP to arrest elongation and the capacity of the SRP receptor to release the arrest might be of important regulatory significance. Modulation of the arrest-releasing activity either by other, as yet unidentified, components or by direct modification of SRP or the SRP receptor, or both, may provide the cell with an on–off switch for translocation-coupled protein synthesis and thereby provide a mechanism for a fast and regulatable response to a variety of physiological stimuli.

Both SRP and the mode of co-translational protein translocation seem to be highly conserved through evolution (Müller *et al.* 1982; Talmadge *et al.* 1980). SRP therefore appears to be an integral and indispensable component of the protein synthesis machinery of living cells assuring the correct topogenesis of a specific subset of proteins (Blobel 1980). Considering its structural features and its intimate (although most likely transient) functional association with ribosomes, it could almost be regarded as a 'third ribosomal subunit' functioning as the adaptor between the cytoplasmic translation and the membrane-bound protein translocation machinery.

REFERENCES

Anderson, D. J., Walter, P. & Blobel, G. 1982 *J. Cell Biol.* **93**, 501–506.
Bielinska, M., Rogers, G., Rucinsky, T. & Boime, I. 1979 *Proc. natn. Acad. Sci. U.S.A.* **76**, 6152–6156.
Blobel, G. 1980 *Proc. natn. Acad. Sci. U.S.A.* **77**, 1496–1500.
Erickson, A. H., Walter, P. & Blobel, G. 1982 (In preparation.)
Gilmore, R., Blobel, G. & Walter, P. 1982*a* *J. Cell Biol.* (Submitted.)
Gilmore, R., Walter, P. & Blobel, G. 1982*b* *J. Cell Biol.* (Submitted.)
Kreibich, G., Freienstein, C. M., Pereyra, B. N., Ulrich, B. L. & Sabatini, D. D. 1978*a* *J. Cell Biol.* **77**, 488–505.
Kreibich, G., Ulrich, B. L. & Sabatini, D. D. 1978*b* *J. Cell Biol.* **77**, 464–487.
Li, W. Y., Reddy, R., Henning, D., Epstein, P. & Busch, H. 1982 *J. biol. Chem.* **257**, 5136–5142.
Marcantonio, E. E., Grebenan, R. C., Sabatini, D. D. & Kreibich, G. 1982 *Eur. J. Biochem.* **124**, 217–222.
Meyer, D. I. & Dobberstein, B. 1980*a* *J. Cell Biol.* **87**, 498–502.
Meyer, D. I. & Dobberstein, B. 1980*b* *J. Cell Biol.* **87**, 503–508.
Meyer, D. I., Krause, E. & Dobberstein, B. 1982*a* *Nature, Lond.* (In the press.)
Meyer, D. I., Louvard, D. & Dobberstein, B. 1982*b* *J. Cell Biol.* **92**, 579–583.
Müller, M., Ibrahimi, I., Chang, C. N., Walter, P. & Blobel, G. 1982 *J. biol. Chem.* (In the press.)
Stoffel, W., Blobel, G. & Walter, P. 1981 *Eur. J. Biochem.* **120**, 519–522.
Talmadge, K., Stahl, S. & Gilbert, W. 1980 *Proc. natn. Acad. Sci. U.S.A.* **77**, 3369–3373.
Ullu, E., Murphy, S. & Melli, M. 1982 *Cell* **29**, 195–201.
Walter, P. & Blobel, G. 1980 *Proc. natn. Acad. Sci. U.S.A.* **77**, 7112–7116.
Walter, P. & Blobel, G. 1981*a* *J. Cell Biol.* **91**, 551–556.
Walter, P. & Blobel, G. 1981*b* *J. Cell Biol.* **91**, 557–561.
Walter, P. & Blobel, G. 1982*a* *Nature, Lond.* (Submitted.)
Walter, P. & Blobel, G. 1982*b* (In preparation.)
Walter, P., Ibrahimi, I. & Blobel, G. 1981 *J. Cell Biol.* **91**, 545–550.
Walter, P., Jackson, R. C., Marcus, M. M., Lingappa, V. R. & Blobel, G. 1979 *Proc. natn. Acad. Sci. U.S.A.* **76**, 1795–1799.
Zieve, G. & Penman, S. 1976 *Cell* **8**, 19–31.

Discussion

P. N. CAMPBELL (*Courtauld Institute, The Middlesex Hospital, London, U.K.*). I should like to ask Professor Blobel's views on the significance of the hydrophobic property of the signal peptide. I realize that it interacts with a hydrophobic SRP but this is not of course an essential feature for the interaction of two proteins.

G. BLOBEL. At this moment we do not know what the precise requirements for the interaction between signal peptide and SRP are. In fact, a direct interaction between the signal peptide and SRP still remains to be demonstrated. This is why we have used the term 'recognition' rather than 'receptor' in naming SRP.

Phil. Trans. R. Soc. Lond. B **300**, 229–235 (1982)
Printed in Great Britain

Recycling of the asialoglycoprotein receptor: biochemical and immunocytochemical evidence

By A. L. Schwartz[1,3], H. J. Geuze[2] and H. F. Lodish[1]

[1]*Department of Biology, Massachusetts Institute of Technology,*
Cambridge, Massachusetts 02139, U.S.A.
[2]*Center for Electronmicroscopy, State University of Utrecht, The Netherlands*
[3]*Division of Pediatric Hematology/Oncology, Children's Hospital Medical Center,*
Sidney Faber Cancer Institute,
Department of Pediatrics, Harvard Medical School, Boston, Massachusetts

[Plate 1]

One of the best documented systems of receptor-mediated endocytosis is the clearance of asialoglycoproteins (ASGP) from the blood plasma by liver parenchymal cells. There are 200000–500000 ligand binding sites per cell, which makes this system favourable for molecular studies of receptor function. By using both biochemical and immunocytochemical approaches, we have obtained evidence for receptor recycling. We have also localized the intracellular site at which the endocytosed receptor and ligand dissociate.

The human hepatoma cell Hep G2 contains abundant ASGP receptors (approximately 225000 per cell). In growing cells approximately 85% of the functional receptors are on the cell surface and the remaining 15% are internal. The maximal rate of ligand uptake in this cell system at 37 °C is approximately 30000 molecules per cell per minute. Each functional receptor can therefore bind and internalize more than 50 ligand molecules during a 6 h period (in the absence of new receptor synthesis), or one ligand each 8 min.

To follow both ligand and receptor during their common endocytosis and to visualize the compartment in which the dissociation of ligand from receptor occurs, we have used our recently developed double-labelling immunocytochemical electron microscopic techniques with purified antibodies against ASGP ligand and ASGP receptor. In normal rat hepatocytes, both ligand and receptor are taken up from the sinusoidal cell surface in clathrin-coated vesicles. Both receptor and ligand are associated with the membrane of small clathrin-coated vesicles close to the cell surface. Larger vesicles, farther removed from the surface, contain ligand accumulated within the lumen. The membranes of these larger vesicles contain little receptor, but receptor was concentrated in detached vesiculotubular extensions, which were largely free of ligand. These vesicles represent the compartment of uncoupling of receptor and ligand (Curl) during their common endocytosis. Ligand contained within the vesicle lumen is then transferred to multivesicular bodies and lysosomes; the tubular extensions may carry receptor back to the cell surface.

Introduction

Many cells are capable of internalizing molecules by receptor-mediated endocytosis (r.m.e.). R.m.e. involves binding of a ligand, such as a hormone, virus, plasma protein, or toxins, to specific receptor molecules functionally exposed at the cell surface. Receptor–ligand complexes generally accumulate in coated pits of the plasma membrane followed by internalization of the complexes in small membranous vesicles (Goldstein *et al.* 1979). Many ligands such as α_2-macroglobulin, low-density lipoprotein and insulin, are transported within membrane-limited compartments to lysosomes, where they are rapidly degraded (Pastan & Willingham

1981). Within acidic endocytic vesicles, the membrane of enveloped viruses fuses with the vesicle membrane, enabling the virus core to reach the cytoplasm (Helenius *et al.* 1980). In many cases, the receptors are spared lysosomal degradation and return to the cell surface.

We and others have been studying the receptor for galactose-terminal carbohydrates of glycoproteins (asialoglycoprotein (ASGP)), which is localized in the hepatic parenchymal cell (Ashwell & Harford 1982). Endocytosis of asialoglycoproteins has been studied in considerable

TABLE 1. SPECIFICITY OF [^{125}I]ASOR† BINDING TO HEPATOMA CELLS

(Dishes were washed and incubated with [^{125}I]ASOR (2 µg ml^{-1}) in the absence or presence of additional agents in the standard manner (see Schwartz *et al.* 1981 *a*). Non-radioactive asialoorosomucoid, orosomucoid and asialogalactoorosomucoid were added at 200 µg ml^{-1}. EDTA was present at 5 mM. Neuraminidase preincubation was performed by incubation with 20 milliunits in 1 ml of phosphate-buffered saline (p.b.s.) for 15 min at room temperature. Results are expressed as the percentage of the total [^{125}I]ASOR bound compared with the control ('none').)

addition	[^{125}I]ASOR bound (%)
none	100
EDTA	13
none/EDTA*	29
asialoorosomucoid	22
orosomucoid	93
asialogalactoorosomucoid	84
neuraminidase preincubation	24

* One set of dishes was allowed to bind [^{125}I]ASOR in the standard manner and was thereafter rinsed and incubated for 5 min at 4 °C with p.b.s. containing 5 mM EDTA.

detail in whole liver *in vivo*, in perfused liver *in situ*, and in isolated rat hepatocytes (Ashwell & Morell 1974; Hubbard *et al.* 1979; Steer & Ashwell 1980; Baenziger & Fiete 1980; Schwartz *et al.* 1980). There are as many as 500 000 high-affinity surface receptors per hepatocyte (Schwartz *et al.* 1980; Zeitlin & Hubbard 1982). In addition, this receptor has been isolated and purified from rabbit, rat and human liver (Hudgin *et al.* 1974; Schwartz *et al.* 1981 *b*). Recent studies have begun to elucidate the characteristics of receptor-mediated endocytosis in this system. Using electron microscopic techniques Hubbard and colleagues have demonstrated the uptake of galactose-terminal glycoproteins by rat hepatic parenchymal cells and followed their subsequent transfer to a series of endocytic vesicles and ultimately to lysosomes (Wall & Hubbard 1981; Zeitlin & Hubbard 1982). Biochemical studies by Tolleshaug *et al.* (1977), Steer & Ashwell (1980) and others have provided evidence for a receptor-mediated uptake of asialoglycoproteins by isolated rat hepatocytes.

Using a combined biochemical and immunocytochemical electron micrographic approach, we have accumulated strong evidence for active recycling of the ASGP receptor and we have precisely localized the intracellular compartment where the dissociation of ASGP from its receptor occurs.

RESULTS AND DISCUSSION

We have performed the biochemical dissection of the pathway for ASGP uptake and degradation in the human hepatoma cell line Hep G2 isolated by Knowles *et al.* (1980). As seen in table 1, Hep G2 cells specifically bind [^{125}I]ASOR.† Such binding studies were per-

† ASOR, asialoorosomucoid.

formed at 4 °C in order to minimize internalization of the ligand. Binding requires the presence of Ca^{2+} and is not substantially effected by the presence of a 100-fold excess (by mass) of orosomucoid or asialoagalactoorosomucoid. Pretreatment of the cells with neuraminidase renders them incapable of binding [125I]ASOR. There are 150000–200000 high-affinity ASOR-binding sites per cell surface. These data are all consistent with the characteristics of

TABLE 2. SPECIFICITY OF RELEASE OF SURFACE-BOUND [125I]ASOR
FROM HEPATOMA CELLS

(Dishes were washed and incubated with [125I]ASOR (2 µg ml⁻¹) in the standard manner (2 h, 4 °C). After washing in p.b.s. containing 1.5 mM $CaCl_2$, the indicated additions were made for the indicated time at 4 °C. Thereafter, one further rinse in p.b.s. with $CaCl_2$ was performed and the samples counted. Results are expressed as a percentage of the total [125I]ASOR bound compared with the control ('none'). (In part adapted from Schwartz et al. (1981a).)

addition (concentration) (time)	[125I]ASOR bound (%)
none	100
EDTA (5 mM) (3 min)	12
N-acetyl-galactosamine	
(100 mM) (10 min)	11
(50 mM) (10 min)	24
galactose (100 mM) (10 min)	62
N-acetyl-glucosamine (100 mM) (10 min)	100
ASOR (200 µg ml⁻¹) (300 min)	95

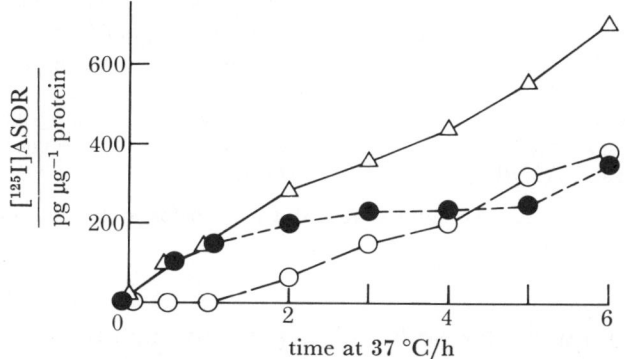

FIGURE 1. Uptake and degradation of [125I]ASOR in human hepatoma cells at 37 °C. Tissue culture dishes containing 10⁶ cells were washed and incubated with [125I]ASOR (2 µg ml⁻¹) for various times at 37 °C. At the appropriate times the media were removed and analysed for 125I degradation products. In addition, the cells were rinsed and analysed for cell-associated [125I]ASOR. The figures represent the mean and range of duplicate determinations of cell-associated radioactivity (●), 125I-degradation products of the media (○) and the sum of the two previous values (△). (Adapted from Schwartz et al. (1982).)

the asialoglycoprotein receptor in rat hepatocytes (Ashwell & Harford 1982). In addition, once bound to its receptor, [125I]ASOR could be readily displaced by either a brief treatment at 4 °C with EDTA or N-acetylgalactosamine but only minimally by galactose. N-Acetyl-glucosamine or ASOR was without effect (table 2). The sensitivity of surface-bound [125I]ASOR to displacement by EDTA or N-acetylgalactosamine provides a sensitive and convenient assay for surface-bound ligand; internalized ligand is resistant to such treatments.

At 37 °C, there is a linear increase in the amount of cell-associated [125I]ASOR during the first 2 h (figure 1). A constant level of cell-associated ligand is reached by 2 h. There is little

[125]I label in degradation products in the medium before 1 h, and the linear increase in ^{125}I-degradation products begins by the second hour (figure 1). As expected, there is no detectable degradation of [^{125}I]ASOR when maintained for 6 h at 37 °C under identical conditions but in the absence of cells (data not shown). As measured by the sum of cell-associated and degraded ^{125}I radioactivity, the overall rate of cellular uptake of ASOR is constant at 0.02–0.03 pmol min^{-1} per 10^6 cells for at least 6 h (figure 1). In 28 independent experiments the rate of [^{125}I]ASOR uptake at a concentration of 2 µg ml^{-1} at 37 °C, assessed over the first 60 min, averaged 0.029 ± 0.001 pmol min^{-1} per 10^6 cells, as we have reported earlier (Schwartz *et al.* 1981 *a*, 1982).

TABLE 3. SURFACE RECEPTOR DISTRIBUTION AND UPTAKE OF [^{125}I]ASOR IN
HEPATOMA CELLS AFTER TRYPSIN TREATMENT

(Dishes containing monolayer cells were rinsed and incubated with [^{125}I]ASOR (2 µg ml^{-1}) at either 4 °C (specific binding) or 37 °C (uptake) (Schwartz *et al.* 1981 *a*). Cell suspensions prepared with either EDTA or EDTA/trypsin were incubated with [^{125}I]ASOR (2 µg ml^{-1}) at either 4 °C (specific binding) or 37 °C (uptake). In addition, samples were incubated with cycloheximide (CLX) (100 µg ml^{-1}) with or without an additional 60 min at 37 °C with ASOR. Results (means ± s.e.) are presented of duplicate (binding) or triplicate (uptake) values. Initially, a fraction x of the functional receptor is internal to the cell, and $(1-x)$ is on the surface. After trypsin treatment there is a residual fraction, T, of surface receptors remaining. Thus, the total remaining fraction of cell-associated receptors is equal to the rate of uptake of [^{125}I]ASOR, relative to control cells. Solution of this equation yields the value of x, the fraction of receptors internal to the cell. (Adapted in part from Schwartz *et al.* (1982).)

treatment	CLX	specific binding pmol/10^6 cells	rate of [^{125}I]ASOR uptake (60 min) pmol/(10^6 cells/min)	receptor on the cell surface (%)
(A) monolayer–EDTA	–	0.29 ± 0.04	0.023 ± 0.001	
suspension–EDTA	–	0.33 ± 0.04	0.023 ± 0.001	
suspension–EDTA/trypsin	–	0.02 ± 0.00	0.004 ± 0.000	87.9
(B) suspension–EDTA/trypsin + ASOR	–	0.04 ± 0.00†	(0.004)	—
suspension–EDTA/trypsin + ASOR	+	0.05 ± 0.01†	(0.004)	75

† These samples were 'stripped' of surface-bound ASOR with EDTA before determination of specific binding with [^{125}I]ASOR.

Because the rate of ligand uptake (ligand flux) is dependent upon the total cell complement of functional receptors that participate in this process, we have determined the cell receptor distribution by destroying cell-surface receptors with protease. As shown in table 3, single-cell suspensions of Hep G2, prepared by treatment of monolayer cultures with an EDTA solution at 4 °C, bind at 4 °C the same amount of [^{125}I]ASOR as do cells assessed under standard conditions in monolayer culture. The rate of ligand uptake at 37 °C is also un-impaired. However, if trypsin is included in the EDTA solution at 4 °C (as is used for dispersing monolayer cells), binding of [^{125}I]ASOR to cells is inhibited by over 90 %, indicating that virtually all surface receptors are destroyed by this protease. When trypsin-treated cells are incubated at 37 °C, uptake of [^{125}I]ASOR is linear with time (data not shown) but is only 20 ± 2 % of that of control cells. Taking into account that only 94 % of the surface receptors is actually destroyed by trypsin, the data in table 3 indicate that in growing Hep G2 cells approximately 88 % of the functional receptor is on the surface, and 12 % is internal. When cells that had been trypsinized were then exposed to cycloheximide to prevent further protein synthesis, and allowed to incubate at 37 °C but in the presence of unlabelled ASOR (2 µg ml^{-1})

for 60 min, the remaining functional receptors redistributed themselves such that approximately 75 % were again on the cell surface and 25 % internal (table 3). Thus, during steady-state ligand binding and internalization, more receptor is internal than in growing cells.

Because the uptake and degradation of ligand continues at a steady rate of 15 000 molecules per cell per minute independent of new receptor synthesis (Schwartz *et al.* 1982) for at least 6 h and because there are 150 000–200 000 binding sites per cell surface, either there must exist a large pool of previously synthesized receptor within the cell, or receptor reuse must occur to some extent. If no reuse occurs, then the functional receptor pool within the cell must be at least 30–60-fold greater than the number of surface receptors. However, in these Hep G2 cells, 86 % of all functional receptors are on the cell surface. As calculated from the total number of functional receptors per cell (225 000) and the rate of ligand uptake (15 000 molecules per minute at an ASOR concentration of 2 µg ml^{-1}), each receptor must recycle the ligand, on the average, every 15.9 min ($= 225\,000 \div 15\,000$ min^{-1}). These observations and calculated values were all obtained at a ligand concentration of 50 nM (2 µg ml^{-1}). Obviously, at higher ligand concentrations the total cycle time will decrease as will the time required for ligand binding, until a point is reached at which binding is no longer rate-limiting. The rate of ligand uptake and degradation at 10–20 µg of [^{125}I]ASOR ml^{-1} (i.e. 30 000 per cell per minute) is double that at 2 µg ml^{-1} (Schwartz *et al.* 1982), and the cycle time at 10–20 µg ml^{-1} is about half that at 2 µg ml^{-1}, or 7.9 min.

At 2 µg ml^{-1} ASOR, binding of ligand to surface receptors requires a mean time of 8.7 min. Internalization of receptor–ligand complexes requires a mean of 2.2 min, whereas a mean of 4.2 min is required for the internalized receptor to dissociate its ligand and return to the cell surface (Schwartz *et al.* 1982). Each of these rate constants was determined by two or more independent means; the sum of these times yields 15.1 min for the total cycle time of the asialoglycoprotein receptor.

ASGP ligands are therefore capable of being taken up and processed through to the lysosomes at a considerable rate (see also Ashwell & Harford 1982), whereas the receptor is apparently spared degradation. Additional biochemical studies have demonstrated that the intracellular half life of ASGP ligand taken up by receptor-mediated endocytosis is about 15–20 min, whereas that of the receptor is probably greater than 40 h (see Ashwell & Harford 1982).

Importantly, in Hep G2 cells, degradation of internalized ligand begins only after 20–30 min, a time much longer than the total cycle time of the receptor. Such studies suggest that receptor is not transferred to lysosomes, a conclusion substantiated by our morphological studies.

We have used our recently developed double-labelling immunocytochemical electron microscopic technique, with antibodies against both ASGP ligand and ASGP receptor, to visualize the compartment in which dissociation of the ligand–receptor complex occurs. Asialofetuin (1–6 mg) was administered in 1 ml physiological saline containing 1.5 mM CaCl$_2$ to adult rats by continuous infusion into a tail vein over 30–60 min, followed by perfusion fixation with 2 % formaldehyde – 0.5 % glutaraldehyde. Cryosectioning and immunolabelling with colloidal gold adsorbed to staphylococcal protein A were essentially as described previously (Geuze *et al.* 1981, 1982a). Affinity-purified monospecific rabbit antibodies against the purified rat liver ASGP receptor (Schwartz *et al.* 1981b) and against purified asialofetuin were employed.

Both ligand and receptor are taken up from the sinusoidal cell surface in clathrin-coated vesicles, which deliver the complexes to vesiculotubular structures (Geuze *et al.* 1982b). Both receptor and ligand were found associated with the membrane of small clathrin-coated vesicles close to the cell surface. Little or no free ligand occurred within the lumen of these vesicles. We also identified other larger vesicles found at some distance from the plasma membrane, which contain ligand accumulated within the lumen. The membranes of these latter vesicles contained little receptor, but receptor was concentrated in detached tubular extensions that were largely free of ligand (figure 2, plate 1). No significant receptor labelling was ever found within the vesicle lumen. Interestingly, receptor was not uniformly distributed along the membrane of these larger vesicles, but was either dispersed in clusters along the vesicle membrane or appeared as accumulations at the poles, where vesicles and thin membranous tubules approximated each other or were continuous. In most such vesicles, receptor labelling was either low compared with the connected or adjacent tubules, or was absent. The intensity of ligand labelling was greatest in larger vesicles, which also showed the lowest receptor labelling.

In the bile capillary region the larger vesicles often contain smaller internal vesicles. These multivesicular bodies lacked receptor but contained abundant ligand in the matrix space outside the internal vesicles.

Therefore, once internalized in coated vesicles, ligand accumulated in vesicles ranging in diameter from 0.2 μm at the sinusoidal cell surface to about 0.8 μm in the lysosomal area. The amount of ligand accumulated and the diameter of the vesicle increased in parallel. In the smaller, more peripheral, vesicles, ligand was membrane-associated, presumably with the membrane-bound receptor that was present in abundance in these vesicles. A diffuse membrane distribution of receptor was seen in those vesicles with receptor associated at ligand. In the vesicles with ligand free in the lumen, receptor was concentrated heavily at those sites of the vesicles at which thin, membranous tubules with intense receptor labelling were connected. This double-labelling pattern strongly suggests that these curl-tailed vesicles represent the compartment of uncoupling of receptor and ligand. We have suggested the acronym CURL to identify this compartment of dissociation (Geuze *et al.* 1982*b*). We believe that these tubules may represent an intermediate in the recycling of the receptor to the plasma membrane.

This study was supported by grant no. GM 27989-03 from the National Institutes of Health. A. L. S. is a John and George Hartford Foundation Fellow.

REFERENCES

Ashwell, G. & Harford, J. 1982 Carbohydrate recognition systems in the liver. *A. Rev. Biochem.* **51**, 431–469.

Ashwell, G. & Morell, A. G. 1974 The role of surface carbohydrates in the hepatic recognition and transport of circulating glycoproteins. *Adv. Enzymol.* **44**, 99–128.

Baenziger, J. U. & Fiete, D. 1980 Galactose and *N*-acetyl-galactosamine-specific endocytosis of glycopeptides by isolated rat hepatocytes. *Cell* **22**, 611–620.

Geuze, H. J., Slot, J. W., Van der Ley, P. A. & Scheffer, R. C. T. 1981 Use of colloidal gold particles in double labeling immunoelectronmicroscopy of ultrathin frozen sections. *J. Cell Biol.* **89**, 653–655.

Geuze, H. J., Slot, J. W., Strous, G. J. A. M., Lodish, H. F. & Schwartz, A. L. 1982*a* Immunocytochemical localization of the receptor for asialoglycoprotein in rat liver cells. *J. Cell Biol.* **92**, 865–870.

Geuze, H. J., Slot, J. W., Strous, G. J. A. M., Lodish, H. F. & Schwartz, A. L. 1982*b* Intracellular site of asialo-glycoprotein receptor–ligand uncoupling. Double-label immunoelectronmicroscopy during receptor mediated endocytosis. (Submitted for publication.)

Goldstein, J. L., Anderson, R. G. W. & Brown, M. S. 1979 Coated pits, coated vesicles, and receptor mediated endocytosis. *Nature, Lond.* **279**, 679–685.

Helenius, A., Kartenbeck, J., Simons, K. & Fries, E. 1980 On the entry of Semliki forest virus into BHK-21 cells. *J. Cell Biol.* **84**, 404–420.

Hudgin, R., Pricer, W. E., Ashwell, G., Stockert, R. & Morell, A. 1974 The isolation and properties of a rabbit liver binding protein specific for asialoglycoprotein. *J. biol. Chem.* **249**, 5536–5543.

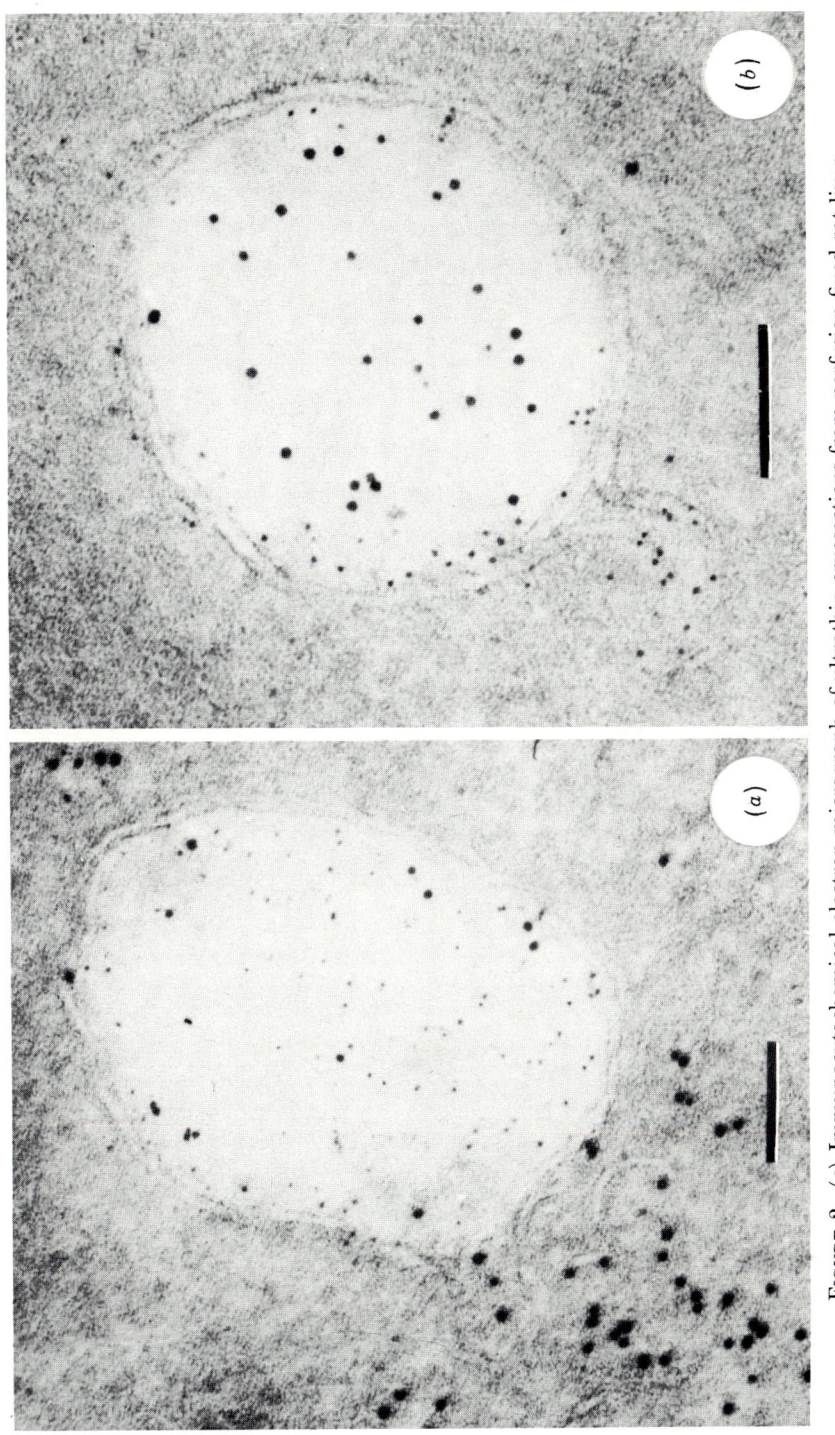

FIGURE 2. (a) Immunocytochemical electron micrograph of ultrathin cryosections from perfusion-fixed rat liver during continuous infusion of asialofetuin. Ligand was labelled first with anti-ASF antibody and then with 5 nm colloidal gold–protein A. Thereafter ASGP receptor was immunolabelled with antibody and then with 8 nm colloidal gold–protein A. Free ligand can be seen in the lumen of the vesicular portion of this sorting vesicle, which also shows scarce and heterogeneous receptor distribution. Receptor labelling is intense over the connecting tubules. Bar = 0.1 μm. (b) Similar to (a) except that receptor is labelled with 5 nm gold whereas ligand is labelled with 8 nm gold. Receptor is located predominantly at the fold where a tubule with heavy receptor labelling is connected. Most of the ligand is present free within the vesicle lumen. Bar = 0.1 μm.

Hubbard, A. L., Wilson, G., Ashwell, G. & Stukenbrok, H. 1979 An electronmicroscope autoradiographic study of the carbohydrate recognition systems in rat liver. *J. Cell Biol.* **83**, 47–64.

Knowles, B. B., Howe, C. C. & Aden, D. P. 1980 Human hepatocellular carcinoma cell lines secrete the major plasma proteins and hepatitis B surface antigen. *Science, Wash.* **209**, 497–499.

Pastan, I. H. & Willingham, M. 1981 Journey to the center of the cell: role of receptosome. *Science, Wash.* **214**, 504–509.

Schwartz, A. L., Fridovich, S. E., Knowles, B. B. & Lodish, H. F. 1981*a* Characterization of the asialoglycoprotein receptor in a continuous hepatoma line. *J. biol. Chem.* **256**, 8878–8881.

Schwartz, A. L., Fridovich, S. E. & Lodish, H. F. 1982 Kinetics of internalization and recycling of the asialoglycoprotein receptor in a hepatoma cell line. *J. biol. Chem.* **257**, 4230–4237.

Schwartz, A. L., Marshak-Rothstein, A., Rupp, D. & Lodish, H. F. 1981*b* Identification and quantification of the rat hepatocyte asialoglycoprotein receptor. *Proc. natn. Acad. Sci. U.S.A.* **78**, 3348–3352.

Schwartz, A. L., Rup, D. & Lodish, H. F. 1980 Difficulties in the quantification of asialoglycoprotein receptors on the rat hepatocyte. *J. biol. Chem.* **255**, 9033–9036.

Steer, C. J. & Ashwell, G. 1980 Studies on a mammalian hepatic binding protein specific for asialoglycoproteins. *J. biol. Chem.* **255**, 3003–3013.

Tolleshaug, H. T., Berg, T., Nilsson, N. & Noren, K. R. 1977 Uptake and degradation of ^{125}I-labelled asialofetuin by isolated rat hepatocytes. *Biochim. biophys. Acta* **499**, 73–84.

Wall, D. A. & Hubbard, A. L. 1981 Galactose-specific recognition system of mammalian liver: receptor distribution on the hepatocyte cell surface. *J. Cell Biol.* **90**, 687–696.

Zeitlin, P. L. & Hubbard, A. L. 1982 Cell surface distribution and intracellular fate of asialoglycoproteins: a morphological and biochemical study of isolated rat hepatocytes and monolayer cultures. *J. Cell Biol.* **92**, 634–647.

Regional Blood Transfusion Centre,
Crescent Drive,
BRENTWOOD, Essex.